SEP - 6 1995

Color Atlas of

DISEASES AND DISORDERS OF THE SHEEP AND GOAT

KARL A. LINKLATER BVM&S, PhD, FRCVS

Director,
Scottish Agricultural College Veterinary Services,
Edinburgh, UK

MARY C. SMITH DVM

Associate Professor of Large Animal Medicine,
Department of Clinical Sciences,
New York State College of Veterinary Medicine,
Cornell University, Ithaca,
New York, USA

N WOLFE

Copyright © 1993 Mosby–Year Book Europe Limited
Published in 1993 by Wolfe Publishing, an imprint of
Mosby–Year Book Europe Limited
Printed by BPCC Hazell Books Ltd, Aylesbury, England
ISBN 0 7234 1708 3

For full details of all Mosby–Year Book Europe Limited titles
please write to Mosby–Year Book Europe Limited, Brook House,
2–16 Torrington Place, London WC1E 7LT, England.

A CIP catalogue record for this book is available from the
British Library.

**Library of Congress Cataloging-in Publication Data has
been applied for.**

Contents

Preface

This atlas is intended to provide a ready and convenient reference text to help in the recognition of the diseases of sheep and goats. Classically, sheep do not exhibit many characteristic external signs of ill health. Furthermore, they are often kept under extensive systems of management, with minimal day-to-day supervision. Goats are frequently observed more closely because of their sociable interactions with people and their use as dairy animals. However, like sheep, they often show vague or nondistinguishing signs, even when seriously ill. Post mortem examination, therefore, plays a dominant role in confirming the presence of many ovine and caprine diseases.

The majority of illustrations are of gross post mortem changes. These are supplemented with a limited number of photomicrographs of histological lesions where appropriate, but care has been taken not to overdo this aspect. Wherever possible, photographs of animals displaying external signs of ill health have been included.

All illustrations are accompanied by short descriptive paragraphs highlighting the features important in reaching a diagnosis. However, the text has been kept to a minimum and does not deal with specialised laboratory diagnostic techniques or treatment and disease prevention.

K. A. Linklater
M. C. Smith

Acknowledgements

Kenneth W. Angus BVMS, DVM, FRCVS

We should like to acknowledge the major contribution made in the production of this atlas by Dr Kenneth W. Angus, formerly of the Animal Diseases Research Association, Moredun Research Institute, who has carried much of the burden of getting the text and illustrations ready for publication.

Contributors to the Atlas

It is a pleasure to acknowledge the contributions of various organisations and individual colleagues, without whose generous donations of colour slides it would have been impossible to produce this Colour Atlas. We do appreciate, however, that it is not always possible for contributors of colour slides to be absolutely positive as to their original source. We have tried our best to acknowledge the source of all materials sent to us. However, it may be that we have inadvertently published material belonging to a colleague who was not approached for permission. If this should be so, then we can only apologise for what we hope will be taken as a bona fide error.

We have been most fortunate to have had free access to several large collections of slides. In this context, we wish to acknowledge the generosity of the Directors of the Animal Diseases Research Association for permission to use many slides from the photographic archives of Moredun Research Institute, Edinburgh. We are greatly indebted to Dr J.M. King, New York State College of Veterinary Medicine, Cornell University, for laying his extensive slide collection at our disposal, and for his helpful comments on the text. The help of the various Veterinary Investigation Centres within the Scottish Agricultural College (SAC) Veterinary Services in providing their collections of slides is also acknowledged.

Many individual colleagues in SAC Veterinary Services provided slides from their personal collections, and we are grateful to Douglas Gray, Barti A. Synge, Jill R. Thomson, A. Ogilvie Mathieson, Alastair Greig, David M.L. Barber, Norman S.M. McLeod, Bill T. Appleyard, George B.B. Mitchell and Harry M. Ross for their contributions. Major contributions were also gratefully received from colleagues at the Moredun Research Institute, including Drs David Buxton, Gareth E. Jones, J. Michael Sharp, David R. Snodgrass, Hugh W. Reid, Frank Jackson, Robert L. Coop, Neville F. Suttle and David A. Blewett, and from Lorna Hay, John S. Gilmour, John A. Spence, David C. Henderson, Edward W. Gray and Alan S. Inglis.

We also wish to thank colleagues in the Ministry of Agriculture, Fisheries and Food (MAFF) Veterinary Investigation Service in England, including Dr David A. Dyson, Andrew Holliman, Marjorie E. Brown and Victor R. Simpson; also Drs Martin Jeffrey and Ranald Munro of the MAFF Lasswade Laboratory, Scotland, and Drs Stuart M. Taylor and David Bryson of the Northern Ireland Department of Agriculture Stormont Laboratory, for valuable contributions.

We are grateful to colleagues at the Centre for Tropical Veterinary Medicine, Edinburgh, including Drs M.M.H. Sewell and Gordon Scott, Archie Hunter and H.R. Urquhart, for providing slides, and to Drs Alex I. Donaldson and Paul Kitching of the Institute for Animal Health, Pirbright Laboratory, England, for similar assistance. Dr Keith Barnett of the Animal Health Trust, Kennet, England, provided the slides of bright blindness, and we are grateful for his help.

Colleagues at many veterinary colleges and universities, in both the UK and abroad, have provided substantial help, and we are happy to acknowledge this here. Ken W. Head, Ian S. Beattie, Drs Steve McOrist, Neil Watt and Gordon H.K. Lawson of the Royal (Dick) School of Veterinary Studies, Edinburgh; Drs Hal Thomson, David J. Taylor and Pauline E. McNeil of Glasgow University Veterinary School; Drs David Lloyd and Tony Wilsmore of the Royal Veterinary College; Dr Agnes C. Winter from Liverpool University Faculty of Veterinary Science, and Nigel P. French, University of Bristol, provided us with many wonderful slides.

Colleagues in Australia have been equally generous, and we are happy to acknowledge the contributions of Professor A.A. Seawright, University of Queensland; Dr Phil W. Ladds, James Cook University of North Queensland; Professor Clive Huxtable, Murdoch University; Professor John R. Egerton, University of Sydney; Dr Ian Beveridge, University of Melbourne; Dr Peter W. Johnson, New South Wales Department of Agriculture, Elizabeth MacArthur Agricultural Institute; Dr Ralph Dowling, Queensland Department of Environment and Heritage; J.R.W. Glastonbury, New South Wales Agriculture and Fisheries, Murray and Riverina Region; and Dr J.R. Allen, Animal Health Laboratories, Western Australia Department of Agriculture.

We have had substantial help from colleagues in Spain, including Dr Lorenzo Gonzalez, Dr Ana Garcia, Dr Ramon A. Juste and Luis A. Cuervo, of Gobierno

Vasco Departmento de Agricultura y Pesca, Servicio de Investigacion y Mejora Agraria, Derio, Bizkaia, and Drs Marcello de las Heras and Jose A. Garcia de Jalon, of the Faculdad de Veterinaria, Universidad de Zaragoza. We are also grateful for contributions from Dr David West, Massey University, Palmerston North, New Zealand, from Dr Martha J. Ulvund, State Veterinary Research Station for Small Ruminants, Norway, from Dr Eitan Rapaport, State of Israel Ministry of Agriculture Veterinary Services and Animal Health, and from Dr Jim DeMartini, Colorado State University.

Other colleagues who provided valuable material were Dr J. Kirkbride, Max Bonniwell, Professor R.M. Barlow, T. Boundy, Dr M.R. Hall of the Natural History Museum, London, Dr R.G. Stevenson, Dr Jim Orr, Western College of Veterinary Medicine, Saskatoon, Drs Neil L.J. Gilmour, Don Bailey, Andrew Eales, W.J. Hartley, G. Dirksen, D.O. Cordes, A. de Lahunta, Mark Riordan, Kevin Morgan, Enrico Lippi Ortolani, São Paulo, Brazil, Ken McEntee, R. Miller, S. Lloyd, W.C. Rebhun, J.J. Bertone, John E. Post, Frank Garry, Cleon Kimberling, C.S.F. Williams, Nancy East, R.I. Coubrough, Koos Coetzer, C. Gall, Parviz Hooshmand-Rad, Professor O. Bwangamoi and Tony Hutson of the Bat Conservation Trust, who kindly provided a slide of a vampire bat (5.11).

We are also grateful to the Editor of *In Practice*, for kindly allowing us to publish numerous photographs which previously had been published in that Journal, and to Blackwell Scientific Publications for allowing us to use material published in *Diseases of Sheep* (Editors I.D. Aitken and W.B. Martin), second edition, 1991. Similarly, we appreciate the permission granted by Coopers Pitman-Moore Ltd to allow publication of photographs of several ectoparasites in the chapter on Skin Diseases. Grateful thanks are due to Dr M. Mathewson for his help in this respect.

Finally, we wish to thank Jackie Calder, Publications and Audiovisual Department, Scottish Agricultural College, and Brian J. Easter, Moredun Research Institute, for photographic services during the preparation of the Atlas.

1 Selected Heritable and Developmental Abnormalities

Congenital defects of sheep and goats are numerous and can be found in every major body system. It is not within the scope of this chapter to illustrate more than a limited selection. Although some have a clearly defined genetic basis, many are the result of environmental factors, such as intrauterine virus infections or toxicities, which interfere with normal fetal development. Some congenital defects are described in other chapters: intersexes and freemartins in Chapter 8; entropion in Chapter 12; and congenital icterus in Chapter 14.

Defects of the cardiovascular system

Transposed aorta

In this condition the aorta instead of the pulmonary artery receives the blood from the right ventricle (**1.1, 1.2**). Where complete transposition occurs, the pulmonary artery emerges from the left ventricle. In some instances, where the aorta receives blood from both ventricles, animals do not survive for long.

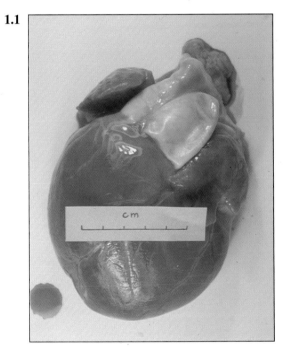

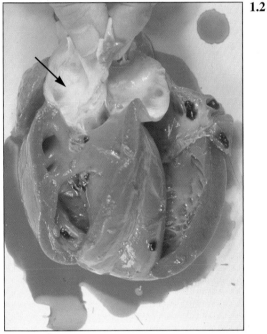

1.1 Transposed aorta in a newborn lamb.

1.2 Transposed aorta. The heart is open to show openings of vessels. Note the aorta and pulmonary artery emerging from the right ventricle (arrow).

1.3

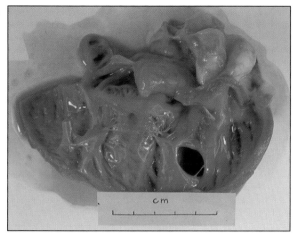

Ventricular septal defects
Closure of the ventricular or atrial septa normally takes place *in utero*. Ventricular septal defects are the most common of all cardiac defects (**1.3**). The defect can involve either the muscular or the membranous portions of the septum. A left-to-right shunt of blood leads to right ventricular hypertrophy. If decompensation occurs, death from heart failure ensues.

1.3 Ventricular septal defect in a lamb.

Defects of the central nervous system

Cerebellar atrophy (daft lamb disease)
This disease is a congenital atrophy of cerebellar components and is believed to be inherited. Affected lambs are unable to rise, and collapse if placed in a standing position. The head is often held in dorsiflexion—'stargazing' (**1.4**). Less severely affected lambs can stand with difficulty and even suck, but have widely spread legs and a waddling gait.

Degenerative changes, including pallor or hypochromasia, vacuolation and, or necrosis, may be seen in the Purkinje cells of the cerebellar cortex (**1.5**), and in neurons in the dentate and other nuclei.

1.4

1.4 Daft lamb disease. Typical star-gazing posture.

1.5

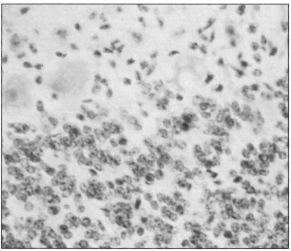

1.5 Daft lamb disease. Degeneration of Purkinje neurons in the cerebellar cortex. Haematoxylin and eosin.

Cerebellar bifurcation

The cause of this simple defect (**1.6**) is unknown.

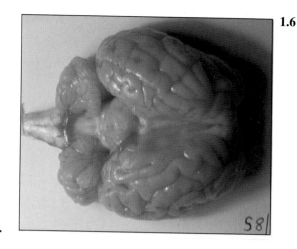

1.6

1.6 Cerebellar bifurcation in a newborn lamb.

Cerebellar hypoplasia

This is a common congenital defect. Many instances are due to intrauterine viral infections. The anatomical changes vary from microscopic lesions only to virtual absence of the cerebellar peduncles (**1.7**).

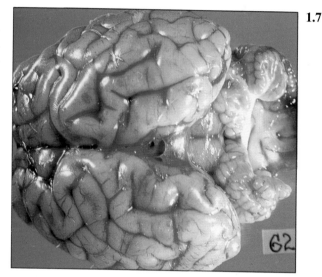

1.7

1.7 Cerebellar hypoplasia in a newborn lamb.

Hydrocephalus

Although this defect is fairly common, the aetiology is seldom clear. The brain deformity is usually internal, though in the instance illustrated there is a dome-like bulging of the skull (**1.8**).

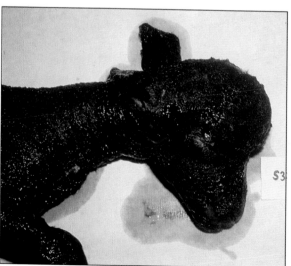

1.8

1.8 Congenital hydrocephalus.

Defects of the axial skeleton

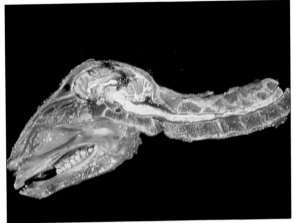

1.9

Atlanto-occipital fusion and subluxation

Atlanto-occipital fusion (**1.9**) is probably caused by congenital absence of the odontoid process of the axis. Affected lambs or kids may be clinically normal at birth, but any stress will result in spinal injury and progressive tetraplegia. Luxation is more common in goats than in lambs, and is due to failure of the odontoid process to fuse with the axis. Clinical signs vary from neck pain to development of tetraplegia over a period of months.

1.9 Atlanto-occipital fusion in a sheep.

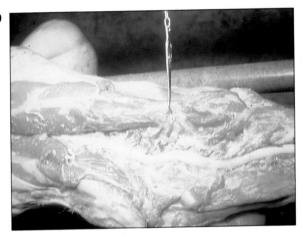

1.10

Scoliosis

'Kinky spine' or scoliosis (**1.10**) is a congenital malformation of the axial skeleton caused by genetic or environmental factors, such as intrauterine viral infections or some plant poisonings. The defect rarely occurs alone, but usually is part of a package of skeletal defects, including arthrogryposis.

1.10 Scoliosis. Note pinching of the spinal cord.

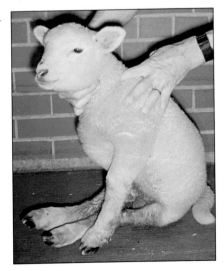

1.11

Spina bifida

Defective closure of the neural tube in this congenital defect is expressed by absence of the dorsal portions of the lumbosacral vertebrae (**1.11, 1.12**). One form of this defect in Icelandic sheep is caused by an autosomal recessive trait.

1.11 Spina bifida. Posterior paresis.

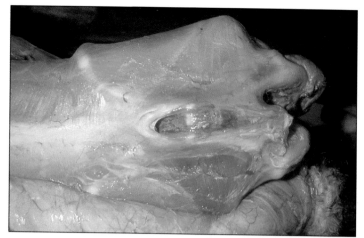

1.12

1.12 Spina bifida. The spinal defect is exposed.

Craniofacial defects

Agnathia

Congenital absence of the mandibles (agnathia, **1.13**) is more common in sheep than any other mammalian species. The defect is usually caused by environmental factors operating during pregnancy. Agnathia is often associated with ear pinna defects; for example, microtia or synotia (fused ears).

Brachygnathia (undershot jaw)

This is one of the most common of all craniofacial defects. Some forms are caused by a recessive gene, though a variety of environmental factors can also cause the defect (**1.14**). If not too severe, mastication is possible, but it is generally desirable to cull affected animals.

1.13

1.13 Agnathia.

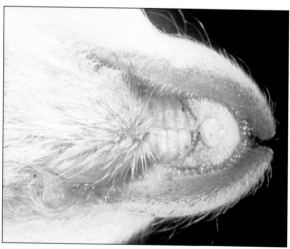

1.14

1.14 Undershot jaw.

1.15

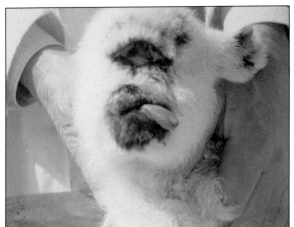

Cyclopia

The plant *Veratrum californicum* (false hellebore), if consumed by the dam on the 14th day of gestation, when the ocular field normally divides into two primordia, induces cyclopia (**1.15**) in lambs and kids. Other teratogenic effects of the plant include absence of the pituitary and malformation of the long bones in the extremities.

1.15 Cyclopia. This stillborn lamb has a single eye and facial deformities resulting from injury to the 14-day-old fetus.

1.16

Nasal bone deviation

This defect rarely occurs in isolation but usually is part of a group of localised craniofacial defects, often lethal. In the sheep illustrated, the defect had no clinical significance (**1.16, 1.17**).

1.16 Nasal bone deviation in a ewe.

1.17

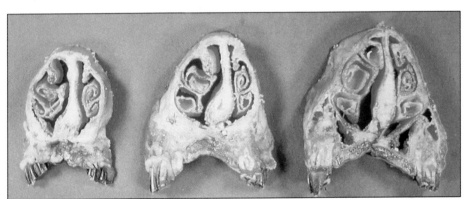

1.17 Nasal bone deviation. Cross-sections of nasal bones at different levels.

Defects of bones, joints and limbs

Achondroplasia

Lambs affected by this genetically controlled condition are abnormally short-legged due to failure of normal development of the cartilaginous growth plate in the long bones. The disease is most common in Merinos, Norwegian sheep, Southdowns, Dorsets, Scottish Blackfaces and Cheviots (**1.18**) and some crosses. Affected animals seldom reach maturity.

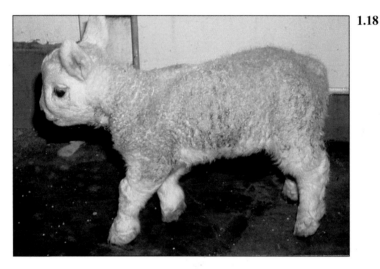

1.18 **1.18 Achondroplastic Cheviot lamb.**

Congenital arthropathy

This is a sporadic defect, in which otherwise apparently normal lambs in Australia are born with ankylosed joints in complete flexion. The long bones are often abnormally slender, and the joints incompletely formed (**1.19**). No particular genetic or environmental factor has been implicated in this defect.

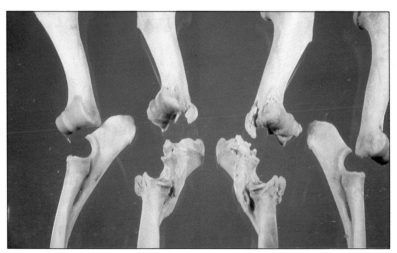

1.19 **1.19 Congenital arthropathy**. Examples of affected bones.

Congenital hypertrophic osteopathy

This rare disease is characterised by marked overgrowth of cortical bone, especially in the jaws and long bones (**1.20, 1.21** and **1.22**). Affected lambs are stunted and walk stiffly. The cause is unknown.

1.20

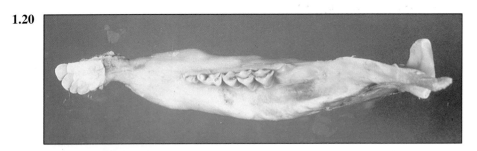

1.20 Ovine hypertrophic osteopathy. Thickened mandibular ramus.

1.21

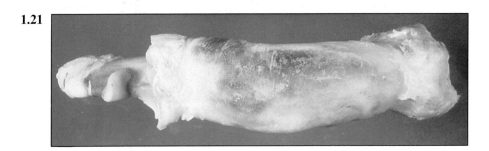

1.21 Ovine hypertrophic osteopathy. Thickened radial shaft.

1.22

1.22 Ovine hypertrophic osteopathy. Tibia split to show cortical bone thickening.

Hereditary chondrodysplasia (spider lamb)

This is an autosomal recessive defect of black-faced breeds, e.g. Suffolks and Hampshires, in North America. Most affected lambs can be identified at birth. Limbs are disproportionately long and spider-like. The most striking defect is medial deviation of the carpus (**1.23, 1.24**) or, less commonly, the tarsus.

Other skeletal deformities include severe scoliosis or kyphosis (**1.25**), concave sternum, severe Roman nose, incomplete development of the anconeal process, degenerative changes in the joints, or arthrogryposis. These changes can be demonstrated radiographically.

1.23

1.24

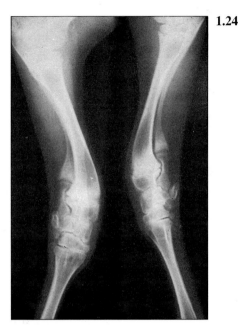

1.23 Spider lamb. Medial deviation of the carpus.

1.24 Spider lamb. Radiograph of malformed forelimbs.

1.25

1.25 Spider lamb. Long limbs and spinal curvature are evident.

Polymelia and polydactyly

Both polymelia (extra limbs, **1.26**) and polydactyly (extra digits) are fairly common skeletal defects, especially in lambs. These are often accompanied by other defects. Environmental factors are usually assumed to be the cause.

Rib deformities

Congenital costal deformities occur in sheep, with narrowing of the thoracic inlet (**1.27**). Rib defects are usually associated with other skeletal defects, such as scoliosis.

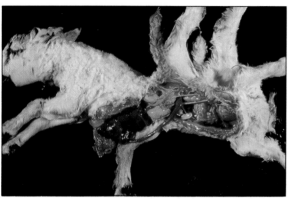

1.26 Polymelia. A seven-legged lamb is shown.

1.27 Congenital rib deformities in a sheep.

Defects of the gastrointestinal tract

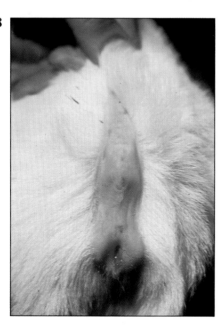

Atresia ani (imperforate anus)

This condition is the most common congenital defect of the lower gastrointestinal tract. A heritable form occurs in certain lines of goats (**1.28**). Sometimes the defect is confined to failure of perforation of the membrane separating the hindgut from the ectodermal anal tissue, and can be corrected surgically. More commonly, the defect involves both anus and rectum. It may be isolated, or associated with malformations of the genitourinary tract (**1.29**). In female lambs or kids, the terminal portion of the rectum may open into the vagina.

1.28 Atresia ani in a kid.

1.29 Atresia ani, bifurcated scrotum and partial hypospadia.

Intestinal aplasia

Aplasia of the ileum is the most common segmental abnormality in the small intestine, but actually is rare in sheep or goats. Although usually complete, it is occasionally incomplete (stenosis). The defect can take various forms: the obstruction may be a simple membrane or the blind ends of the gut may be separated by some distance (**1.30**). The use of contrast media demonstrates the lesion clearly (**1.31**). The portion of

1.30 Ileal aplasia. Appearance at necropsy.

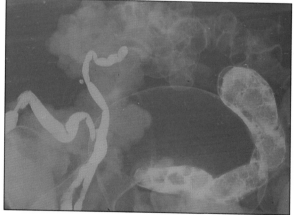

1.31 Ileal aplasia in a lamb, demonstrated on X-ray using contrast medium.

1.32

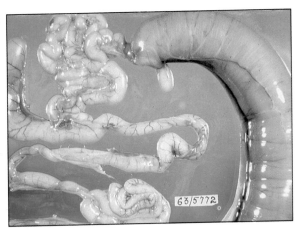

1.32 Intestinal obstruction. Dilated bowel proximal to obstruction.

gut proximal to the obstruction is widely dilated (**1.32**), while that distal to it is empty.

1.33

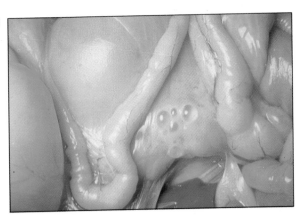

1.33 Pancreatic cysts in a young lamb.

Pancreatic cysts

Cysts are occasionally found in the pancreas of lambs (**1.33**) and have no significance unless there are associated cysts in other organs, e.g. liver or kidney (see Cystic/polycystic kidneys, page 23).

Defects of musculature

Diaphragmatic hernia

Most diaphragmatic hernias occur as a result of incomplete fusion of the opposing pleuroperitoneal folds, and are found in the dorsal, tendinous portion of the diaphragm. They vary in size, but substantial ones permit the herniation of considerable volumes of abdominal viscera into the thorax (**1.34**). Trauma can contribute to their occurrence.

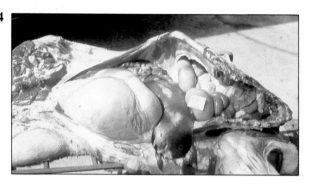

1.34 Diaphragmatic hernia. Abdominal organs in the thorax.

Hanging head of Suffolks

This is a congenital deficiency of neck musculature found in some Suffolk flocks. The defect is seldom detectable until animals are mature, when affected sheep obviously cannot raise their heads fully (**1.35**). It should not be confused with a form of nutritional weakness of neck musculature which can occur in newborn Suffolk lambs, and which responds well to supplementary feeding.

1.35 Hanging head. The Suffolk ewe nearest the camera cannot raise her head properly.

Schistosomus reflexus

This gruesome defect, in which the viscera appear to surround the skeleton and skin (**1.36**), is an extreme form of open abdominal hernia integrated with a variety of skeletal defects. The cause is unknown.

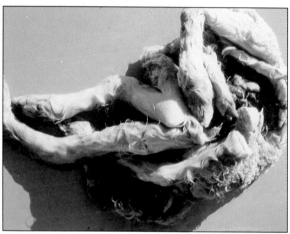

1.36 Schistosomus reflexus.

Umbilical hernia

Open umbilical hernias are usually congenital. They can result in eventration of loops of gut through the patent umbilical orifice (**1.37**). These hernias can be reduced successfully by surgery.

1.37 Umbilical hernia with eventration.

Reproductive system defects

Amorphus globosus

This unusual fetal malformation is an extreme form of arrested development, though in this specimen from a goat mature skin and hair are present (**1.38**)

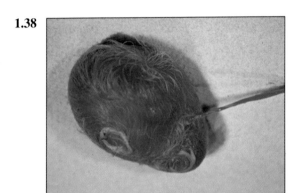

1.38 Amorphus globosus goat fetus.

Bifurcated scrotum

This simple defect represents incomplete fusion of the primordia, and may be isolated or accompanied by other defects, e.g. hypospadia, or incomplete fusion of the urethra (see **1.29**). Breeders often discriminate against this trait, but it appears to favour testicular cooling and thus fertility in hot climates.

Double teats

Supernumerary teats completely separate from the main teats occur occasionally in does but are rarer in ewes. They have no clinical importance, but both double and fused teats (**1.39**) interfere with milking.

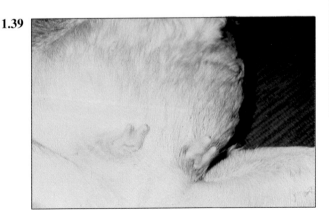

1.39 Double teats in a doe.

Inguinal hernia, scrotal hernia

Scrotal hernia is merely an exaggerated version of inguinal hernia and is found in both bucks and rams. The viscera pass down the inguinal canal and may lie in the cavity of the tunica vaginalis (**1.40**). The diameter of the internal inguinal ring and the tendency to herniation is inherited. In scrotal hernia, there may be testicular degeneration. There is a risk of eventration if castration should be contemplated (see also Chapter 8).

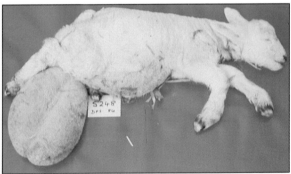

1.40 Scrotal hernia.

Siamese twins

Conjoined twins (**1.41**) are monozygotic twins that are imperfectly formed. They are not uncommon in sheep, but rare in goats. Ageing of ova is said to be a common cause.

1.41 Conjoined twin lambs.

Skin defects

Congenital hypotrichosis

This can occur in both sheep and goats. The exact genetic basis is unknown but the trait is probably recessive. Affected Dorset sheep are born with hairless areas on the face and lower legs (**1.42**).

1.42

1.42 Congenital hypotrichosis in a Dorset ewe.

Congenital wattle cysts

Wattle cysts are hereditary (probably dominant) in goats and may be unilateral or bilateral (**1.43**). The cyst is usually present at birth in a subcutaneous location attached to the base of a wattle. The cyst may enlarge with time as it fills with thick or thin clear fluid. Excision may be requested to avoid confusion with caseous lymphadenitis abscesses.

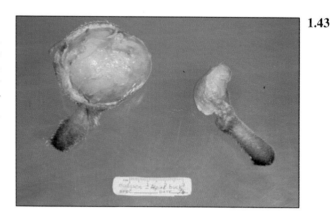

1.43

1.43 Congenital wattle cysts in a goat.

Dermatosparaxis (easy tear)

This disease is an hereditary collagen dysplasia seen in Dala lambs in Norway and Border Leicester–Southdown crosses in Australia. The defect results from deficiency of an enzyme, *N*-terminal procollagen peptidase, causing an excess of low-tensile procollagen in the skin. The skin of affected lambs is abnormally fragile and lacerates easily (**1.44**). Even the trauma of sucking may cause skin tearing and oral lesions that fail to heal and become infected.

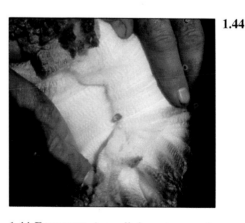

1.44

1.44 Easy tear. A small dermatosparaxis wound in a lamb.

Epidermolysis bullosa

This fatal genetically controlled defect occurs in Suffolk and Southdown sheep in New Zealand. Subepidermal clefts develop beneath the basal layer within a few hours of birth, with vesicle formation leading to erosions over pressure points (**1.45**), lips and nasolabial plane. Oral lesions include erosions in the tongue, gingiva and hard palate (**1.46**). Lesions may also develop at the coronary band.

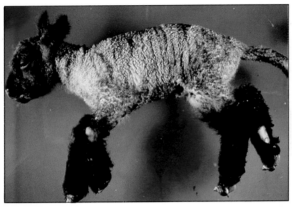

1.45 Epidermolysis bullosa. Skin lesions on lower limbs.

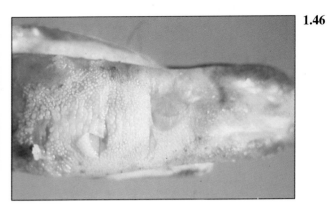

1.46 Epidermolysis bullosa. Tongue lesion.

Epitheliogenesis imperfecta

This rare condition probably has a genetic cause. Sharply demarcated, variably sized defects in the epithelium are present at birth (**1.47**) and the underlying tissues soon become traumatised and infected.

1.47 Epitheliogenesis imperfecta in a lamb.

Redfoot

This disease is found mainly in the Scottish Blackface and Welsh Mountain breeds. Some sources consider redfoot to be a form of epitheliogenesis imperfecta, but it is best to regard it as a dermoepithelial separation. It results in a variety of lesions in lambs only a few days old. Redfoot is probably hereditary.

Typically, there is shedding of the horn of the hooves, exposing the underlying sensitive corium and causing severe lameness (**1.48**). Blisters or ulcers may be seen also in the mouth, nose, inner aspects of the ears and coronary band, while the horn buds may be shed. Such lambs should be destroyed and the sire culled.

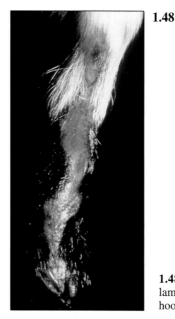

1.48

1.48 Redfoot. A Blackface lamb with a skin defect and hoof loss.

Urinary tract defects

Congenital dilatation of the renal pelvis

In this condition, the renal cortex is narrowed and the renal pelvis widely dilated at birth (**1.49**). The cause may be ureteric obstruction *in utero*.

Cystic/polycystic kidneys

Congenital renal cysts are the most common of all congenital renal deformities (**1.50**). They may be associated with cysts in other organs, e.g. liver or pancreas. Affected lambs are usually stillborn or die within a few weeks of birth of renal failure. A genetic cause has been suggested, but the nature of the trait is unknown.

1.49

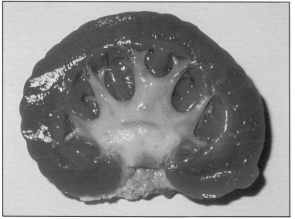

1.49 Congenital renal pelvis dilatation in a lamb kidney.

1.50

1.50 Polycystic kidney.

2 Generalised Infectious Diseases Found Worldwide

This chapter contains a number of viral and bacterial infections that are not confined to tropical or sub-tropical regions of the world or utilise vectors that are highly dependent upon environmental conditions for survival. The list is not comprehensive, and does not include some diseases that involve more than one body system but predominate in one system. For example, caseous lymphadenitis is dealt with in Chapter 14 while caprine arthritis–encephalitis (CAE) is discussed in several chapters, as clinical signs in an individual are usually limited to one system.

Anthrax

Anthrax is caused by *Bacillus anthracis,* a gram-positive spore-bearing bacillus. The vegetative form of the bacillus has a limited capacity to survive. The spores, which form on exposure to oxygen, however, are extremely durable and can survive in the soil for many years.

Sheep and goats probably become infected by ingesting the spores of the bacillus, and sheep also can become infected via the skin. Spores pass to regional nodes and undergo vegetative multiplication. Rapidly dividing organisms quickly spread along the lymphatics, swamping the local defences successively until they gain access to the blood. Death quickly follows as a result of a massive septicaemia.

Diagnosis is established by demonstrating the characteristic bacilli in blood smears (**2.1**). Carcases of animals that have died of anthrax decompose very rapidly; thus, the sample should be taken from the ear or tail tip, where putrefactive processes may be delayed. Necropsy should be avoided if anthrax is suspected, in order to avoid sporulation as well as human infection. Necropsy findings include multiple haemorrhages in mucous and serous membranes, and a tarry appearance of the blood. Splenomegaly is not an invariable finding, as in cattle.

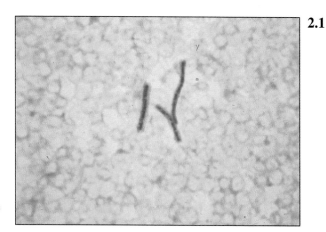

2.1

2.1 Anthrax organisms in sheep blood. The bacilli have squared ends and a distinctive capsule. Polychromatic stain.

Black disease

Infectious necrotic hepatitis (black disease) is caused by *Clostridium novyi*. Spores of this organism can enter and live in the liver, mainly of sheep, where they can lie latent for many months. These spores are harmless unless anaerobic conditions are created in a non-immune animal, in which case the spores germinate and the organism assumes its vegetative form, releasing potent lethal, necrotising and haemolytic toxins.

Anaerobic conditions occur during the migration of immature liver flukes, usually *Fasciola hepatica* and *Dicrocoelium dendriticum*. The cestode *Cysticercus tenuicollis* occasionally acts as a precipitating cause.

Death occurs rapidly and without warning signs. The reflected skin is dark, due to marked venous congestion, giving the disease its name. There may be oedema of the sternal subcutis. Body cavities contain considerable amounts of yellow serous fluid, which clots rapidly on exposure to air. One or several light-coloured rectangular or irregular areas of necrosis, 2 to 3 cm in diameter and surrounded by zones of hyperaemia, are found characteristically in the liver (**2.2**). Haemorrhagic tracks caused by fluke migration are also present. Diagnosis can be made by direct immunofluorescence of an impression smear from the edge of the lesion.

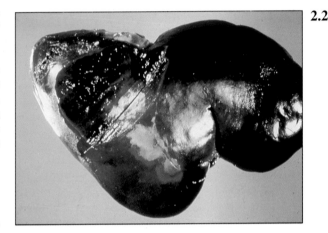

2.2

2.2 Black disease. A focal necrotic lesion in the liver.

Bluetongue

Bluetongue is an arthropod-borne viral infection that affects both sheep and goats, though sheep tend to be more seriously affected. The cause, a virus of the genus Orbivirus in the family Reoviridae, exists as numerous serotypes. The virus is transmitted mainly by biting midges (*Culicoides* spp.).

The disease causes fever, lameness (**2.3**), and excessive salivation with nasal and ocular discharges (**2.4**). Initially, hyperaemia of the lips, buccal (**2.5**) and nasal mucosa, and conjunctivae are prominent. A few days later mouth lesions develop, consisting of shallow ulcers that affect the lips, tongue, buccal mucosa and

2.3

2.3 Bluetongue. Lameness and tiptoe stance in the acute disease.

2.4

2.4 Bluetongue. Facial oedema (R) and nasal discharge (L).

gums. Crusts form around the nostrils (**2.6**) and these, plus the oral lesions, can be confused with orf (see Chapter 10). The face swells and thus photosensitisation (Chapter 10) should be considered in differential diagnosis. Lameness is due to inflammation of the coronary band (**2.7**) and laminitis. Wool break can occur if the whole skin is hyperaemic. Haemorrhages are often present in the heart (**2.8**) or rumen pillars. Diagnosis is by virus isolation or serology.

Infection of pregnant ewes at 4 to 8 weeks' gestation can cause abortion or deformity of the fetus, with hydranencephaly, porencephaly and/or, retinal dysplasia (see Chapter 9).

2.5

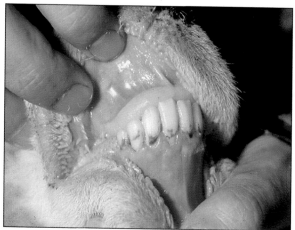

2.5 Bluetongue. Hyperaemia of the oral mucous membranes.

2.6

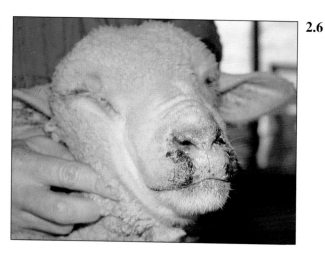

2.6 Bluetongue. Crusts round the eyes and nares.

2.7

2.7 Bluetongue. Lesions at the coronary band.

2.8

2.8 Bluetongue. Haemorrhage at the base of the pulmonary artery.

2.9

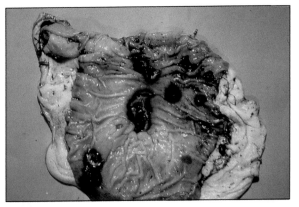

2.9 Braxy. Focal haemorrhagic necrosis in the abomasum.

2.10

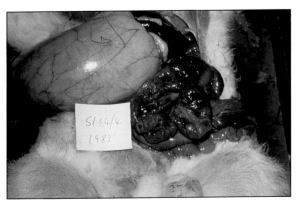

2.10 Struck. Localised enterotoxaemia in the small intestine.

2.11

2.11 Pulpy kidney. Excess pericardial fluid.

Braxy

This disease, which is caused by the anaerobic spore-forming organism *Clostridium septicum* and was once a common cause of death in sheep of all ages in cool climates, is now rare due to vaccine control. At necropsy, the abdominal cavity contains blood-tinged fluid, and the serosal surface of the abomasum may be partially coated by fibrin. Areas of the abomasal mucosa may be red and haemorrhagic, or necrotic (**2.9**). Gelatinous oedema and emphysema are usually present in the submucosa. A fluorescent antibody test is used to identify the organism.

Enterotoxaemias

Apart from lamb dysentery, these diseases are caused by toxins of *Clostridium perfringens* type C (struck) and type D (pulpy kidney disease of lambs, enterotoxaemia of older sheep). Types B and C also cause a haemorrhagic enteritis in very young lambs. The spores of all these organisms are present in the soil, and develop into the toxin-producing vegetative forms only under anaerobic conditions in the gut.

Struck was originally found in the Romney Marsh district of Kent, England. Caused by rapid multiplication of *C. perfringens* type C with elaboration of alpha and beta toxins, the clinical course is ultra-short and rapidly fatal. The peritoneal cavity usually contains a large amount of pale yellow or pink fluid. The small intestines are hyperaemic (**2.10**) and may be ulcerated. There are excess pleural and pericardial fluid and subepicardial haemorrhages.

In pulpy kidney disease, strong lambs 4 to 10 weeks of age or fattening stock 6 to 12 months old are principally affected. Few are seen alive, and these may show various neurological signs, stertorous breathing and collapse. The cause is rapid multiplication of *C. perfringens* type D in the gut with elaboration of a prototoxin which is converted to epsilon toxin by the action of trypsin. At necropsy the animal is in prime condition. There are pale yellow fluid and fibrin in the pericardial sac (**2.11**), and subepicardial haemorrhages. Intestinal contents are fluid, and the intestines congested (**2.12**) The kidneys are swollen and autolyse rapidly (**2.13**); if held under a gentle stream of running water, the parenchyma is washed away, leaving only the cottony cortical framework. Diagnosis is by toxin neutralisation in laboratory mice or by ELISA.

Enterotoxaemia is somewhat different in goats. The disease is more commonly associated with a sudden change in diet than with a continuous fattening ration. Diarrhoea is a prominent clinical finding, although peracute cases may not live long enough to show diarrhoea.

Abdominal pain can be a feature of peracute cases.

Pericardial effusion occurs as in sheep, but there is often a striking necrotising colitis. A chronic form of enterotoxaemia has been recognised in mature goats. These animals show a progressive weight loss with intermittent episodes of loose faeces. This condition must be differentiated from coccidiosis, lactic acidosis, salmonellosis and paratuberculosis (Johne's disease).

2.12 Congested intestines in pulpy kidney disease.

2.13 Pulpy kidney. Necrotic 'cotton-wool' kidney cortex.

Colibacillosis

Coliform bacteria are normal inhabitants of the intestine of all lambs and kids. Both invasive and non-invasive strains of *Escherichia coli* can occur, the latter being mostly harmless commensals. However, there are certain non-invasive isolates (enterotoxigenic *E. coli*, or ETEC) which possess an antigenic pilus (K99+ adherence factor, **2.14**) that enables them to adhere closely to the enterocytes in the small intestine, though only during the first 24-36 hours of life (**2.15**). Such isolates produce an enterotoxin, stable toxin (ST), that acts on adenyl cyclase, the enzyme cascade within the enterocyte responsible for fluid uptake, by reversing its action so that fluid pours from the cells into the gut lumen, causing acute diarrhoea.

Animals orally infected by ETEC are usually under two days old. They have acute watery diarrhoea, and rapidly become dehydrated, prostrate, cold and comatose. Many will die unless treated by oral rehydration. Diagnosis is by culture, followed by identification of the K99+ adherence factor by slide agglutination.

Invasive isolates of *E. coli* sporadically cause septicaemias, meningitis or joint infections in lambs or kids (see Chapter 5), and some highly virulent strains can cause a necrotising abomasitis (see Chapter 4). Infection usually takes place through the navel. Poor colostral immunity is an important predisposing factor (see Watery mouth, page 37).

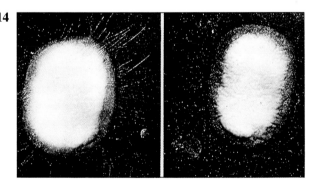

2.14 Enterotoxigenic *E. coli* with adhesive pili (left).

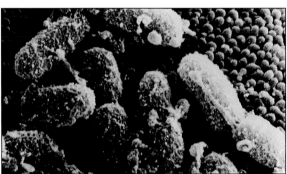

2.15 Enterotoxigenic *E. coli* adhering to microvilli.

Foot-and-mouth disease

This highly contagious disease is caused by Aphthoviruses of the family Picornaviridae, and is a major cause of economic loss in cattle, swine, and to a lesser extent sheep and goats. Acute lameness, salivation and nasal discharge are the first signs noticed in an infected flock or herd (**2.16**). Affected animals are pyrexic. Initially, there is blanching of the dental pad, and there may be vesicles or erosions in the mouth, dental pad, gums, lips (**2.17**), tongue, interdigital space, round the bulbs of the heels or along the coronary band (**2.18**). Vesicles may also occur on the teats, vulva or prepuce.

Rupture of vesicles in the mouth or interdigital space (**2.19**) leads to secondary infection, with necrotic mouth ulcers and footrot being common sequelae. Affected ewes may develop mastitis, while affected rams may refuse to serve. At necropsy, in addition to the lesions in the mouth and feet, ulcers may be found in the rumen (**2.20**). Sudden deaths in young lambs result from direct action of the virus on the heart muscle, which may show erythematous mottling or striping (**2.21**).

Diagnosis is made by virus isolation from vesicular fluid, or by the use of an ELISA test to detect antigen in vesicular epithelium collected from the mouth.

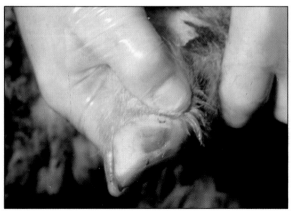

2.16 Foot-and-mouth disease. Acute lameness.

2.17 Foot-and-mouth disease. Lip erosions and blanching of the dental pad.

2.18 Foot-and-mouth disease. Vesicle on the coronary band.

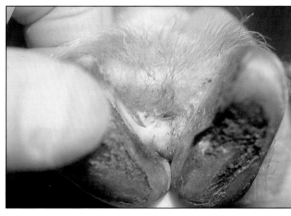

2.19 Foot-and-mouth disease. Ruptured vesicles in the interdigital space.

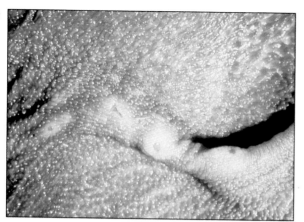

2.20

2.20 Foot-and-mouth disease. Erosions in the rumen.

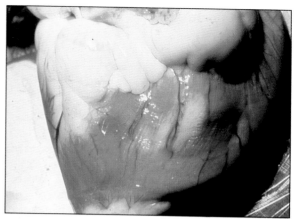

2.21

2.21 Foot-and-mouth disease. Tiger-striping of the myocardium.

Johne's disease (paratuberculosis)

The cause of this disease is infection with *Mycobacterium johnei (M. paratuberculosis)*, a gram-positive, acid-fast organism transmitted in the faeces. Sheep can be infected with both pigmented and non-pigmented forms. The latter behave in culture just like those isolated from cattle, but pigmented strains grow very slowly and sparsely in artificial media.

Infection of sheep and goats causes slow, progressive wasting over many months (**2.22**), but there is often no diarrhoea. Infection with pigmented strains in sheep causes a chronic ileitis and colitis, with bright

yellow pigmentation of the mucosa (**2.23, 2.24**) that results from colonisation with masses of the pigmented bacteria. The ileum may be thickened, with a velvety or granular appearance.

Afferent lymphatic vessels in the mesentery may be thickened and convoluted and may contain white caseous or calcified nodules (**2.25**). Mesenteric nodes are invariably enlarged and may contain whitish gritty nodules when cut open (**2.26**). Microscopy of the gut shows focal or diffuse accumulations of epithelioid macrophages and multinucleated Langhans' cells.

2.22

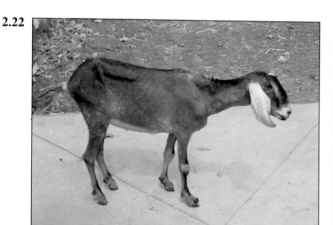

2.22 Johne's disease. An emaciated Nubian goat.

2.23

2.23 Johne's disease. Sheep ileum infected with a pigmented strain.

2.24 Johne's disease. Sheep caecum infected with a pigmented strain.

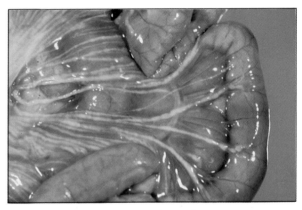

2.25 Johne's disease. Chronic lymphangitis in the mesentery.

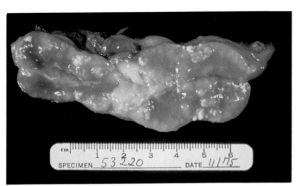

2.26 Johne's disease. Calcified lesions in an enlarged mesenteric lymph node.

2.27 Johne's disease. Acid-fast bacilli in ileal mucosa Ziehl–Neelsen stain..

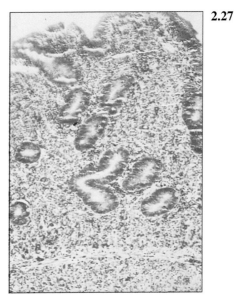

2.28 Johne's disease. Complication by chronic salmonellosis.

Acid-fast organisms can be identified in these lesions, but are most numerous in infections with pigmented strains (**2.27**). Crypt atrophy occurs late in the disease. Caseous or, more probably, calcified nodules may be found in Peyer's patches and mesenteric nodes.

Diagnosis is based on histological findings, on recognition of the typical acid-fast bacteria in smears of ileal mucosa or mesenteric node stained by Ziehl-Neelsen's method, or by culture. Johne's disease may be complicated by other enteric infections, such as salmonellosis (**2.28**).

Lamb dysentery

This affects strong lambs of under two weeks of age, and is caused by the beta and epsilon toxins of *Clostridium perfringens* type B. The clinical disease lasts only a few hours. Affected lambs are in great pain, stop sucking and soon collapse and die. There may be a semi-fluid yellowish or blood-tinged diarrhoea (**2.29**). The principal finding at necropsy is focal haemorrhagic enteritis in the jejunum (**2.30**). Widespread discrete-to-confluent ulceration of the mucosa can be discerned through the serous surface as yellowish areas surrounded by haemorrhagic margins. There may be gas formation in the gut mucosa. Gut contents are a mixture of blood and frothy yellow fluid. The mesenteric lymph nodes are oedematous. The liver may be pale and friable, the kidneys swollen and the lungs congested. Diagnosis is by demonstration of the specific toxins by means of a neutralisation test on laboratory mice or by ELISA. Lambs can be passively protected by colostrum from vaccinated ewes, until old enough to be vaccinated themselves.

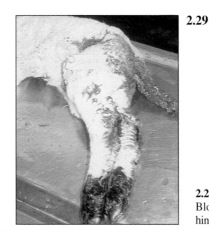

2.29

2.29 Lamb dysentery. Blood-stained tail and hindquarters.

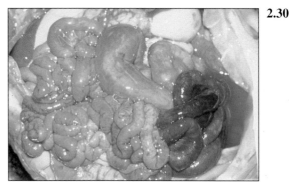

2.30

2.30 Lamb dysentery. Acute enteritis.

Necrobacillosis

Infection of the navel with the filamentous bacterium *Fusobacterium necrophorum* can cause multiple necrotic foci in the liver and occasionally the lungs of young lambs. Affected animals are dull, reluctant to suck and have a hunched posture. Pressure behind the xiphisternum often evinces pain. Morbidity and mortality can be high.

F. necrophorum is a normal inhabitant of the rumen in adult sheep and goats. Under conditions which produce ruminal acidosis, e.g. excessive grain feeding, a ruminitis can develop. This can provide a portal for *Fusobacterium* to invade the rumen wall and pass to the liver, causing hepatic necrobacillosis, especially in sheep.

Fusobacterium produces potent leukocidins, haemolysins and cytoplasmic toxins which cause rapid death of local tissues. The lesions of necrobacillosis are raised brownish circular areas of coagulative necrosis, several centimetres in diameter, in liver and lungs (**2.31, 2.32**), surrounded by zones of hyperaemia and with an offensive odour. In animals which survive, the lesions usually progress to abscess formation.

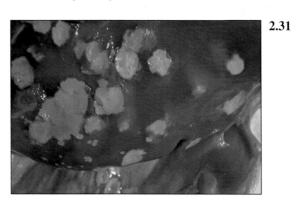

2.31

2.31 Necrobacillosis. Focal liver lesions.

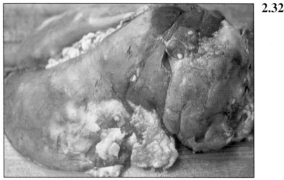

2.32

2.32 Necrobacillosis and *Actinomyces pyogenes* abscesses in the lungs.

2.33

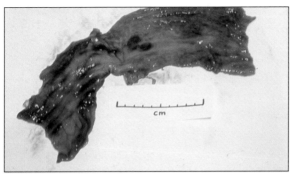

2.33 Salmonellosis. Haemorrhagic enteritis.

2.34

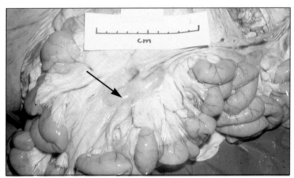

2.34 Salmonellosis. Enlarged posterior mesenteric lymph node (arrow) from which *Salmonella* organisms can be cultivated late in the course of the disease.

2.35

2.35 Salmonellosis. Engorged liver with distension of the gall bladder.

Salmonellosis

Members of the *Salmonella* group of Enterobacteriaceae are short, aerobic gram-negative rods which are differentiated from other members of that family by their biochemical and serological reactions. Few are host-specific, the exception being *Salmonella abortus-ovis*, which causes abortion in sheep.

The most important species pathogenic for sheep are *S. typhimurium*, *S. dublin*, *S. enteritidis*, *S. arizonae* and *S. montevideo*. Outbreaks of salmonellosis in sheep are not common but can be severe. Salmonellosis is rarely a problem in goats.

Many factors can predispose to outbreaks of salmonellosis, not least being in contact with infected calves, feed or water supplies. Infections with *S. abortus-ovis* or *S. montevideo* cause abortion with little clinical disease in adult ewes, though in-contact lambs may have diarrhoea. *S. dublin* and *S. typhimurium* cause high fever and acute, greenish diarrhoea. Many adults as well as lambs can die of septicaemia or progressive dehydration. Pregnant ewes often abort.

Necropsy findings include acute abomasitis and enteritis (**2.33**), with swelling and oedema of mesenteric lymph nodes (**2.34**). Congestion and enlargement of the spleen are common, the liver may be swollen and friable, and the gall bladder distended with dark bile (**2.35**).

Diagnosis can be made by direct culture of faeces, organs and mesenteric lymph nodes. Initial culture in deoxycholate agar or selenite broth will prevent growth of coliforms and other Enterobacteriaciae, which would swamp the developing *Salmonella* colonies by their rapid growth. Typical non-lactose fermenting colonies are then subcultured on deoxycholate citrate agar or MacConkey's medium.

Systemic (septicaemic) pasteurellosis

This disease of weaned lambs or hoggets is a common cause of outbreaks of sudden death in the autumn, particulary where there has been an abrupt change of pasture, feed or other management practices. The cause is infection with *Pasteurella haemolytica* biotype T strains, which differ from biotype A strains in their ability to ferment trehalose as opposed to arabinose in selective media.

Necropsy findings are those of an acute septicaemia. Widespread vascular engorgement with ecchymotic haemorrhages is found in the subcutis of the neck and thorax (**2.36**). The lungs are usually swollen, oedematous and have focal haemorrhages, but are not consolidated. Necrosis of the mucosa is often found in the pharynx, tonsillar crypts, oesophagus (**2.37**) and, occasionally, the abomasum (**2.38**). The liver is often congested and may contain multiple, small (0.5-5.0 mm) miliary lesions of necrosis (**2.39**), and splenic infarcts may be present (see Chapter 14).

In addition to the obvious areas of necrosis, microscopic lesions can be found in the lungs, liver (**2.40**), spleen, adrenal and, occasionally, kidneys. These lesions consist of emboli of *P. haemolytica* surrounded by narrow necrotic zones containing masses of flattened, basophilic leucocytes (oat cells). Meningeal serum protein leakage and choroidal mononuclear cell infiltrates have been reported.

Diagnosis depends on the isolation of large numbers (at least one million colony-forming units per gram of tissue) of *P. haemolytica* T biotype on culture.

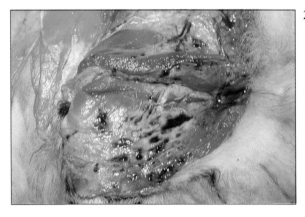

2.36 Systemic pasteurellosis. Engorgement of subcutaneous vessels with focal haemorrhage.

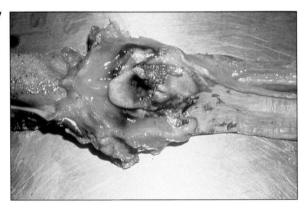

2.37 Systemic pasteurellosis. Pharyngeal and oesophageal necrosis.

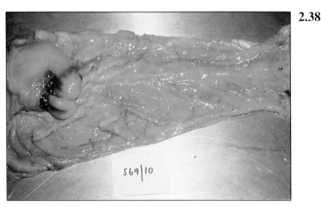

2.38 Systemic pasteurellosis. Acute abomasal ulceration.

2.39 Systemic pasteurellosis. Miliary foci in liver.

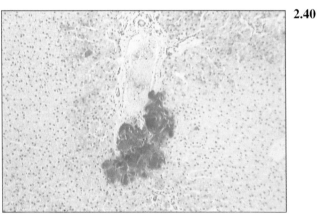

2.40 Systemic pasteurellosis. Embolic lesion in liver. Haematoxylin and eosin.

Tick pyaemia

This disease results from the introduction into the bloodstream of young lambs, or occasionally kids, of the skin bacterium *Staphylococcus aureus* by the bites of feeding ixodid ticks. Bacteria lodge in a variety of sites, including the liver, joints, lungs, kidneys and spinal cord, where multiplication leads to pyogenic infection with abscess formation (**2.41, 2.42**). Tick-borne fever (see Chapter 14) serves to lower the resistance of some lambs to infection with staphylococcus.

Affected animals are covered with feeding ticks and are dull, reluctant to suck, and often lame or paraplegic due to spinal abscesses compressing the spinal cord.

Morbidity is high, and many young lambs die or have to be culled later. At necropsy, multiple abscesses and suppurative arthritis are the usual findings. Careful search may be needed to find spinal abscesses.

2.41 Tick pyaemia. Abscesses in the spine, joint, heart, spleen and liver.

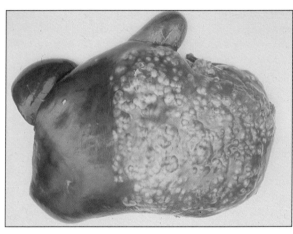

2.42 Tick pyaemia. Multiple liver abscesses.

Tuberculosis

Both sheep and goats are susceptible to tuberculosis, though the disease is fairly uncommon in these species. *Mycobacterium bovis* or, less commonly, *M. avium*, is the cause. In goats, infection is probably by the respiratory route, as lesions are usually mainly in the thorax. In sheep, both respiratory and alimentary routes of infection may be seen. Pulmonary infections become disseminated via the air passages within the lungs and thence into the regional lymph nodes, and then to a variety of other organs as bacilli enter the bloodstream by means of the thoracic duct. Alimentary tract infections cause multiple lesions in the mesenteric lymph-node chain.

Typical necropsy findings are multiple yellow caseous or calcified nodules in a variety of tissues and regional lymph nodes (**2.43, 2.44** and **2.45**). Lesions usually contain typical giant cells, and acid-fast bacilli can readily be demonstrated in sections or smears from the lesions using Ziehl-Neelsen's method.

2.43 Tuberculosis. Miliary liver tubercles.

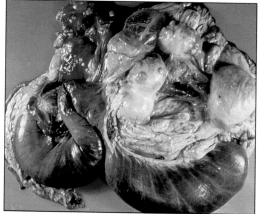

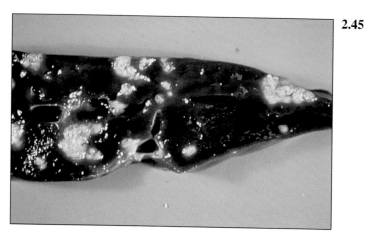

2.44 Tuberculosis. Lesions in the mesenteric lymph nodes.

2.45 Tuberculosis. Cross-section of an affected liver.

Watery mouth

This important disease of neonatal lambs can cause heavy losses in affected flocks. The cause is multifactorial: insufficient colostrum intake coupled with infection by strains of *E. coli* that induce endotoxic effects, and reduced gut motility, are probably the most important factors. Affected lambs are 12–72 hours old and are dull and reluctant to suck. Salivation is commonly seen **(2.46)** and lambs have abdominal tympany **(2.47)** with exaggerated gurgling or splashing sounds on ballottement ('rattle belly'). Diarrhoea may also be present. At necropsy, the abomasum is distended with gas or large volumes of saliva **(2.48)**, and there may be a mild enteritis. The meconium may be retained.

2.46 Watery mouth Salivation and depression.

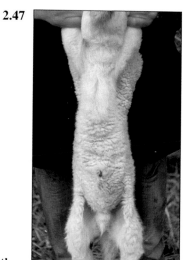

2.47 Watery mouth. Abdominal distension.

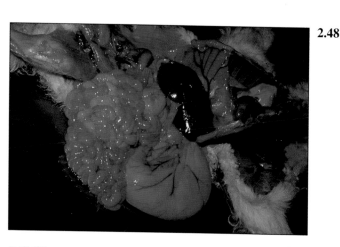

2.48 Watery mouth. Inflamed intestines with frothy contents and bloated abomasum containing milk.

2.49

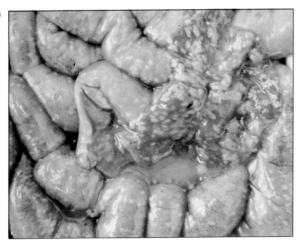

2.49 Yersiniosis. Necrotic and diphtheritic colitis in a goat.

Yersiniosis

Infection with *Yersinia enterocolitica* can cause sudden death without diarrhoea, or death preceded by diarrhoea, resulting from an acute catarrhal enteritis in goats. Young kids are particularly susceptible. Some strains of *Y. enterocolitica* are enterotoxigenic and/or invasive. Invasion and necrosis occur after colonisation of the ileal or caecal lymphoid tissue (**2.49**).

Occasionally, *Y. pseudotuberculosis* can cause enteritis, with multifocal hepatic necrosis and splenitis following oral infection, in sheep.

3 Endoparasitic Diseases of the Alimentary System

Alimentary tract parasites are a constant threat to the survival, growth and productivity of sheep and goats throughout the world. Protozoa, nematodes, trematodes and cestodes abound everywhere small ruminants are maintained, and recognition by means of clinical examination and necropsy can provide early indicators of the nature and scale of a problem. This chapter includes many of the important parasites of the alimentary tract found worldwide, and illustrates the lesions produced by them in their various habitats.

Diseases caused by protozoa

Besnoitiosis of the abomasum and gut

Cysts containing bradyzoites of the Apicomplexan parasite *Besnoitia besnoiti*, which has felidae as its definitive host, are sometimes found in the abomasal and intestinal mucosa of sheep and goats in Iran (**3.1, 3.2**), though cattle are the normal intermediate hosts.

3.1

3.1 *Besnoitia* cysts in a sheep abomasum.

3.2
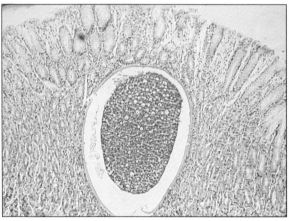
3.2 *Besnoitia* schizont in the colon of a sheep. Trichrome stain.

Coccidiosis

All sheep and goat coccidia are members of the Eimeriidae family and are monoxenous; that is, they have a single host. Infection is by ingestion of an environmentally resistant oocyst that has been passed in the faeces and has sporulated in the stable or on the pasture under suitable conditions of moisture and temperature. The various coccidial species tend to be quite host-specific, even in such closely related mammalian species as sheep and goats.

Eleven species of *Eimeria* are known to infect sheep but only *E. crandallis* and *E. ovinoidalis* are important pathogens. In goats, the important analogous species are *E. ninakohlyakimovae, E. christenseni* and *E. arloingi*. All species have a sexual phase in the intestine, and it is during this part of the life cycle that pathological changes occur. Young animals are most vulnerable to coccidial infections, which cause anorexia, abdominal pain, unthriftiness and diarrhoea, often with straining. The diarrhoea is often bloody if *E. ovinoidalis* is the predominant infection.

Nodular lesions of coccidiosis are often seen in the upper jejunum as small mushroom-like growths (**3.3, 3.4**). In sheep, these formations are the result of infection with *E. ovina*, and have little pathogenic significance.

E. crandallis, E. ovinoidalis and *E. ninakohlyaki-movae* undergo their development in the ileum, while *E. ovinoidalis* will also infect the caecum and colon. *E. christenseni* and *E. arloingi* tend to infect the mid-jejunal region. Motile sporozoites are released from the ingested oocysts in the upper intestine and the asexual stages or meronts (schizonts) may be seen in the jejunum as tiny white spots or plaques in the mucosa. Similar spots in the abomasum are also associated with large meronts in sheep (**3.5**) and goats (**3.6**) but the species of coccidium responsible, though termed *Eimeria* or *Globidium gilruthi*, has not yet been been identified with certainty. Sporozoites occasionally pass via the lymphatic system to develop into meronts in the mesenteric lymph nodes or other sites, but these stages have pathological or clinical significance only if infections are very heavy.

Sexual development (**3.7, 3.8**) is attended by violent effects at the site of infection, as the gut-lining cells, within which the parasites reside, rupture at the end of the cycle to release the oocysts into the gut lumen. Whole sheets of cells slough off into the lumen, with

3.3 Coccidiosis. *Eimeria arloingi* infection in a goat.

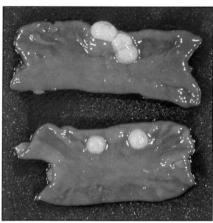

3.4 Coccidiosis. *Eimeria ovina* lesions in a sheep.

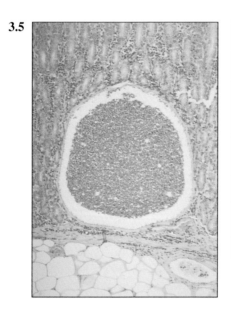

3.5 *Eimeria gilruthi* in a sheep abomasum. Haematoxylin and eosin.

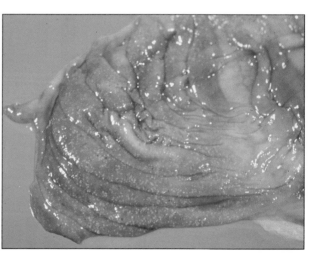

3.6 *Eimeria gilruthi*. Macroschizonts in a goat abomasum.

3.7

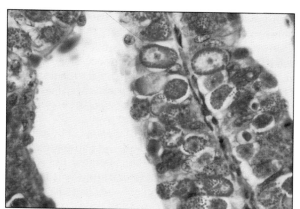

3.8

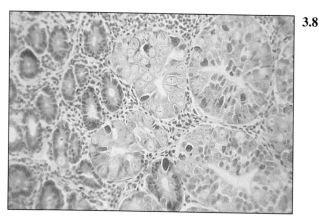

3.7 Coccidiosis. Sexual stages of coccidia in a goat intestine. Trichrome stain.

3.8 Coccidiosis. Sexual stages of *Eimeria ovinoidalis* in a sheep intestine. Haematoxylin and eosin.

discharge of blood cells and proteins. The damaged gut becomes vulnerable at this stage to secondary infections. The villous structure is lost, and large segments of the gut become flattened and non-functional. The crypts are also damaged so that there is no reserve of epithelial cells to restore function.

Diagnosis on the basis of oocyst counts alone is misleading; the flock should be assessed as a whole,

taking account of clinical signs, weight of infection and identification of the coccidial species involved, if possible. Sporulated *E. crandallis* oocysts have broad sporocysts and a shallow polar cap, while those of *E. ovinoidalis* lack a polar cap, have thin walls and their sporocysts are elongated. Accurate diagnosis is complicated by the presence of oocysts of non-pathogenic species.

Cryptosporidiosis

Cryptosporidiosis is caused by *Cryptosporidium parvum*, a coccidian parasite related to the Eimeriidae. The parasite has a direct life cycle, and infection is by faecal–oral transmission. The site of the parasite is mainly the small intestine, and the endogenous stages develop just under the cell membrane of the enterocytes (**3.9**), causing stunting and fusion of the intestinal

villi (**3.10**) and depletion of enzymes such as disaccharidases. This in turn causes diarrhoea. Lambs and kids become infected when only a few days old, and an acute diarrhoea with inappetence may last for over a week. A non-diarrhoeic form of cryptosporidiosis in kids is associated with inappetence and emaciation, sometimes leading to death.

3.9

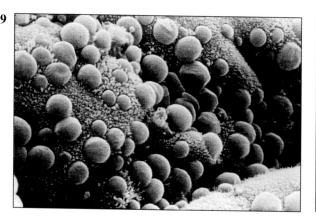

3.10

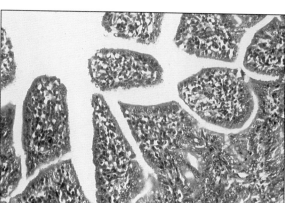

3.9 Cryptosporidiosis. Scanning electron micrograph of *Cryptosporidium parvum* stages in a lamb intestine.

3.10 Cryptosporidiosis. An infected lamb intestine. Cryptosporidia are just visible along the borders of the villi. Haematoxylin and eosin.

3.11

3.11 Cryptosporidiosis. Oocysts in faeces stained by a cold Ziehl–Neelsen method.

Diagnosis of cryptosporidiosis is made by identification of the tiny (4.5 μm) oocysts in faeces (**3.11**). These are fully sporulated in the intestine, and thus are infective as soon as they are excreted into the environment. Up to 10 million oocysts per gram of infected faeces may be passed; thus, environmental contamination can build up rapidly. This can result in heavy mortality in late-born lambs. Oocysts can be demonstrated using a modified acid-fast stain, as shown in **3.11**, or by using Giemsa's stain. The former method is preferred by most laboratories because the oocysts can be recognised using a dry high-power objective. Oocysts fluoresce when stained with phenol-auramine, a useful adjunct if numerous samples need to be scanned.

Diseases caused by trematodes

Dicrocoelium dendriticum infection

This fluke, the lancet fluke, is common in Europe and Asia. The mature fluke is slender and measures only 0.5–1.0 cm long (**3.12**). The fluke has two intermediate hosts. Various genera of land snails are hosts to two generations of sporocysts. The second generation give rise to cercariae, which are expelled from the snail's lung in slime balls. These become infective if swallowed by the ant *Formica fusca*, where they encyst to await ingestion by the definitive host.

Migrating young flukes, often numbering thousands, cause innumerable fine haemorrhagic tracks in the liver (**3.13, 3.14**), while mature flukes cause cholangiohepatitis, with biliary cirrhosis and scarring of much of the liver. Diagnosis is made by detection of the eggs in faeces (**3.15**).

3.12

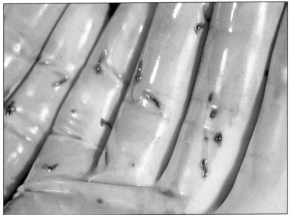

3.12 *Dicrocoelium dendriticum.* Flukes containing ingested blood.

3.1

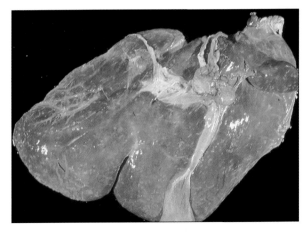

3.13 *Dicrocoelium dendriticum.* Liver damage in a sheep.

14

3.14 *Dicrocoelium dendriticum.* Close-up showing haemorrhagic liver lesions.

3.15

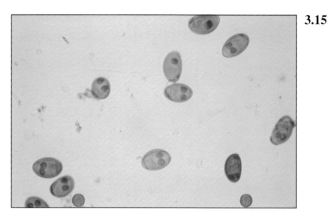

3.15 *Dicrocoelium dendriticum* eggs in faeces.

Fasciola hepatica infection

Fascioliasis is a disease caused by infection with liver flukes (Class Trematoda, sub-order Digenea, family Fasciolidae). *Fasciola hepatica* is the most important and widespread of the group, though in Africa, the Far East and Hawaii it is replaced by *F. gigantica*, which has a comparable life cycle.

F. hepatica has an intermediate host, the water snail *Lymnaea truncatula* or related species. Under appropriate conditions of light and temperature, ciliated miracidia develop in eggs passed in the faeces of infected sheep or goats, and actively infect the snail. After two further stages (the sporocyst and redia stages), the final stages in the snail, the cercariae, are produced and emerge from the snail to swarm up blades of grass or other vegetation (**3.16**), where they encyst to await consumption by the animal.

The cercariae now excyst in the intestine, penetrate the gut wall, and migrate through the peritoneum to the liver. Here they eat their way through the parenchyma (**3.17**) until they reach the bile ducts. During migration, the friable liver may rupture, releasing the young flukes into the peritoneal cavity, with consequent haemorrhage and fatal peritonitis (see Chapter 4).

Once established in the bile ducts, the flukes grow to maturity (**3.18**), and begin to lay eggs some 10–12 weeks after the original infection. Chronic cholangitis results in great thickening and fibrosis (**3.19**), and many affected livers are condemned at meat inspection. Chronic anaemia with regenerative haemopoiesis in the bone marrow accompanies chronic disease (**3.20**). Diagnosis is based on detection of numerous fluke eggs in faeces (**3.21**).

3.16

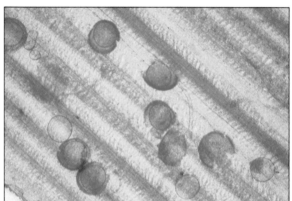

3.16 Fascioliasis. Cercariae of *Fasciola hepatica* on grass.

3.17

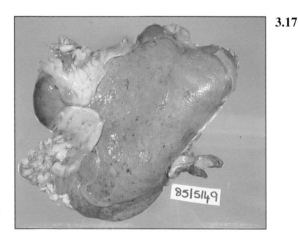

3.17 Fascioliasis. Acute lesions in a sheep liver.

3.18

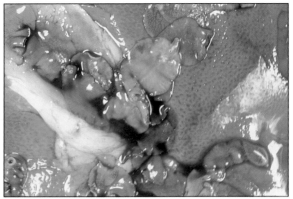

3.18 Fascioliasis. Mature flukes in the bile ducts.

3.1

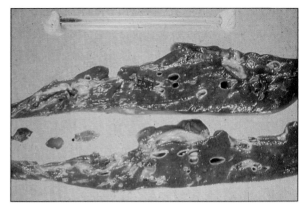

3.19 Fascioliasis. Chronic fibrous thickening of the bile ducts.

3.20

3.20 Fascioliasis. Bone-marrow regeneration in chronic parasitism. Note the low haematocrit shown.

3.2

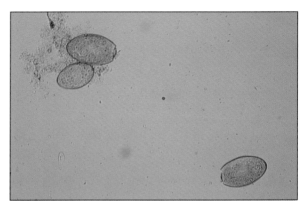

3.21 Fascioliasis. Bile-stained ova in faeces, showing the characteristic operculum.

3.22

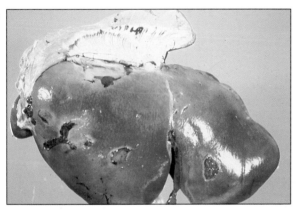

3.22 *Fascioloides magna.* Lesions in a sheep liver.

Fascioloides magna infection

The large liver fluke of North America, *Fascioloides magna*, lives in the liver parenchyma, not in the bile ducts. Its life cycle is similar to that of *F. hepatica*. The young flukes do immense destruction as they wander in the parenchyma, forming tortuous black tracks (**3.22**). A few flukes can kill a sheep.

Paramphistomes

Several genera of *Paramphistomum, Cotylophoron cotylophorum, Gastrothylax crumenifer* and *Fischoederius cobboldi* can be found in the forestomachs of sheep or goats in warm–tropical regions. Metacercariae of these flukes encysted on herbage, when ingested, give rise to immature flukes which inhabit the duodenum (**3.23**), causing severe enteritis and even death.

At necropsy, the duodenal mucosa is thickened, corrugated and covered with immature paramphistomes, which are only a few millimetres long. The flukes embed themselves in the mucosa. There is villous atrophy, crypt elongation and erosion. As they mature, the flukes migrate forwards through the abomasum to the rumen. The adult flukes do little harm.

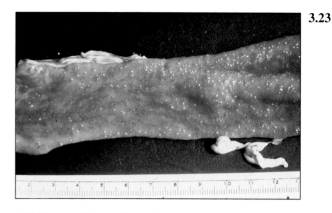

3.23

3.23 Paramphistome flukes in a sheep duodenum.

Diseases caused by nematodes

Haemonchosis

Acute infection with the abomasal worm *Haemonchus contortus* (the 'barber's pole' worm) causes acute haemorrhagic anaemia (**3.24**), with pallor of the mucous membranes, submandibular oedema ('bottle jaw', **3.25**), hyperpnoea and tachycardia. Affected animals are weak and lethargic but there may be no diarrhoea. Haematocrits may be as low as 0.10–0.12 litres per litre (normal 0.30–0.34 litres per litre).

The carcase is usually pallid and often shows ascites and hydropericardium. The blood is usually watery, and the liver pale and friable. The abomasum may contain fresh or altered blood from the lacerations caused by the oral lancet of the adult worm (**3.26**). Abomasal plicae

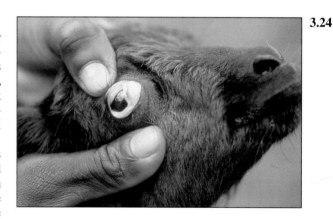

3.24

3.24 Haemonchosis. Severe anaemia in acute infection.

3.25

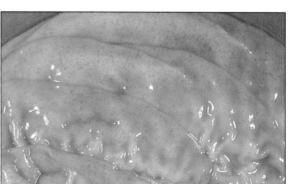

3.26

3.25 Haemonchosis. A sheep with submandibular oedema ('bottle jaw').

3.26 Haemonchosis. Lancet wounds in the abomasal fold.

may be inflamed and turgid, owing to oedema or hypertrophy of the mucosa. Numerous adult *Haemonchus* can usually be seen on the mucosal surface (**3.27**).

In chronic haemonchosis, caused by continuous low-level intakes of larvae, wasting leading to emaciation occurs (**3.28**) with anaemia and hypoproteinaemia.

Affected animals fail to grow and have an open, dull fleece. Apart from the local findings in the abomasum, depletion of red cells leads to stimulation of the bone marrow, which becomes dark red in place of the normal pale fatty appearance (see Chapter 14).

3.27

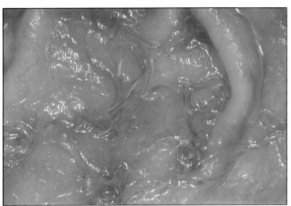

3.28

3.27 Haemonchosis. Adult *Haemonchus contortus* in the abomasum. Note the 'barber's pole' appearance.

3.28 Haemonchosis. Loss of body fat in chronic parasitism.

Ostertagiasis

Ostertagia (now *Teladorsagia*) *circumcincta* (**3.29**) is one of the most economically significant nematode parasites of sheep and goats. Acute infections occur in the early summer, and cause severe abomasitis in grazing lambs, with diarrhoea and unthriftiness. As the larvae undergo development in the abomasal glands, damage to the parietal, or acid-producing, cells results in elevation of the abomasal pH, while pepsinogen from damaged

peptic cells can be detected in the blood. This has greater diagnostic value than faecal egg counts. Compensatory hypertrophy of the mucin-secreting cells in the glands causes a marked thickening of the abomasal plicae.

With the onset of autumn, the larvae of *Ostertagia* undergo arrested development in the mucosal glands (**3.30**), with the formation of nodules in the plicae (**3.31**). In these sites, the parasite can exist for months, to

3.29

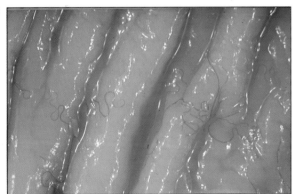

3.30

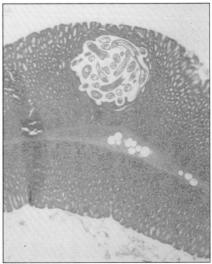

3.29 *Ostertagia circumcincta.* Adult worms in the abomasum.

3.30 *Ostertagia circumcincta.* Arrested larvae in the abomasal mucosa. Haematoxylin and eosin.

emerge in the spring, with the consequent distribution of eggs on the pasture. Sheep can be infected subclinically with *O. ostertagi*, the species of *Ostertagia* that infects cattle, and may act as a reservoir of infection for cattle.

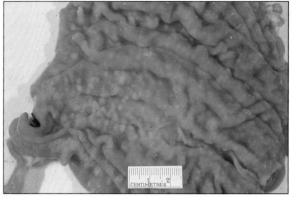

3.31

3.31 *Ostertagia circumcincta.* Nodular hypertrophy and mucosal thickening in chronic parasitism.

Hookworm infection

The species of hookworm that infect sheep and goats are *Bunostomum trigonocephalum*, and, particularly in India, Southeast Asia and Africa, *Gaigeria pachyscelis*. Larvae hatch on the ground and penetrate the skin, passing to the lungs and trachea, and thence to the digestive tract, where they injure the mucosa and feed on blood. They cause anaemia and hypoproteinaemia in young animals, and heavy infections can be fatal.

At necropsy, blood spots and lacerations can be found in the distal small intestine, associated with the presence of the adult worms **(3.32)**.

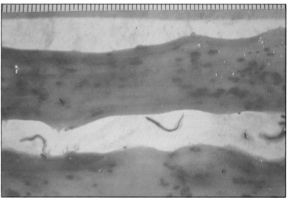

3.32

3.32 Hookworm. *Bunostomum trigonocephalum* infection with lacerations in a goat intestine.

Miscellaneous large intestinal nematodes

Trichuris ovis **(3.33)** is one of the most common worms found in the large bowel and is easily identified by its very long thin neck, which gives it a whip-like appearance. Infections are usually mild and seldom pathogenic. *Chabertia ovina* occasionally causes localised enteritis and mild oedema, with small haemorrhages in the colon. *Oesophagostomum venulosum* scoops out tiny plugs of tissue, and infection can cause minor ulceration in the large intestine.

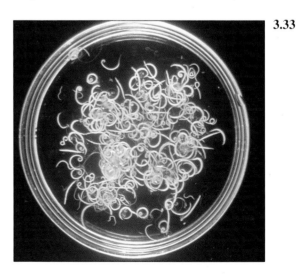

3.33

3.33 *Trichuris ovis.* Adult worms have elongated narrow neck regions.

Nematodiriasis

Of the two species of *Nematodirus* that infect lambs, *N. battus* (**3.34**) and *N. filicollis* (**3.35**), the former is the more pathogenic. The usual pattern of infection is an outbreak of acute diarrhoea (**3.36**), inappetence and lethargy, with some deaths, in lambs in the late spring. Autumn infections have been reported in older lambs, but adult sheep are resistant. Affected lambs lose weight rapidly and can become seriously dehydrated.

Masses of coiled adult parasites and developing larvae may be found in the intestine at necropsy, though they may be expelled terminally. Infection is associated with catarrhal enteritis (**3.37**). Worm eggs, which are often very numerous and of a relatively large size, can survive freezing and thawing in winter, and hatching with release of the infective L3 larvae occurs spontaneously after a cold spell in response to a sudden rise of temperature to above 10°C. There is evidence that young dairy cattle can harbour *Nematodirus*, and contribute to pasture contamination.

3.34

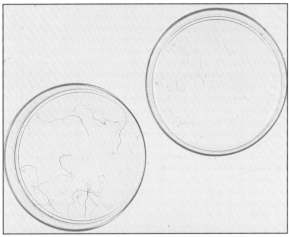

3.34 Size comparison between *Nematodirus battus* (left) and *Cooperia* spp. from sheep.

3.35

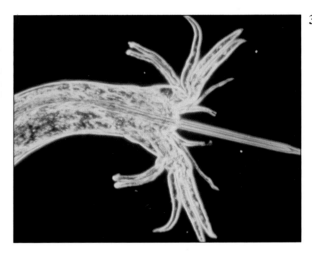

3.35 *Nematodirus filicollis* bursa and spicule.

3.36

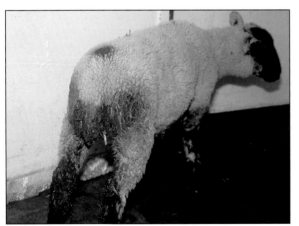

3.36 Nematodiriasis. Acute diarrhoea.

3.37

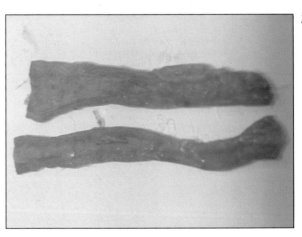

3.37 Nematodiriasis. Acute enteritis.

Oesophagostomum spp. infection

Two species, *Oesophagostomum columbianum* and *O. venulosum*, infect the colon of sheep and goats, though the former has greater economic importance. The third-stage infective larvae penetrate the colonic mucosa and submucosa, and a proportion undergo a second, histotropic phase in the submucosa. Granulomatous nodules, which project from the serous surface and may be calcified, form where these stages have migrated; hence the name 'pimply gut' (**3.38**). Similar nodules may be found in the mesentery and mesenteric nodes (**3.39**), and even the liver, lungs or heart (see Chapter 13). Adult worms cause chronic mucoid colitis, diarrhoea and unthriftiness.

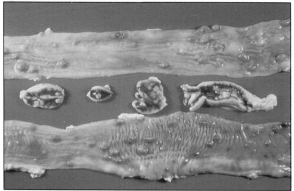

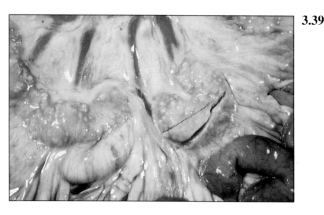

3.38 *Oesophagostomum columbianum.* Worm nodules in 'pimply gut'.

3.39 *Oesophagostomum columbianum.* Worm nodules in mesenteric nodes and intestine.

Trichostrongylosis

Tiny (9–16 mm long) intestinal hairworms (*trichos* = a hair) can cause severe damage to the duodenum and proximal jejunum of grazing lambs during the late summer. In temperate climates, *Trichostrongylus vitrinus* is the main culprit, while its cogener *T. colubriformis* is more commonly found in the southern hemisphere.

Both larval and adult stages burrow into the surface lining of the gut and occupy these channels as holdfasts (**3.40, 3.41**) while they utilise the ingesta in competition with the host. The overall effect is to cause widespread destruction and flattening of the infected portion of the gut (**3.42, 3.43**). Heavy infections—up to 30,000 or so worms—can cause acute diarrhoea and unthriftiness. More insidiously, moderate subclinical infections can

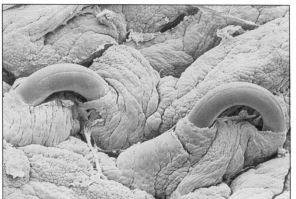

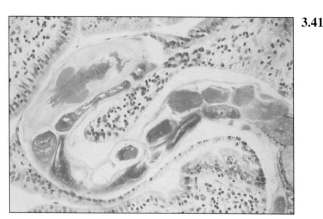

3.40 *Trichostrongylus vitrinus.* Worms burrowing into the mucosa.

3.41 *Trichostrongylus colubriformis.* A worm in the subepithelial channel. Haematoxylin and eosin.

occur, which cause long-term unthriftiness and result in delayed growth and fattening of store lambs. Acute infections can be diagnosed by counting faecal worm egg outputs, but these are low in subclinical infections and therefore of little diagnostic significance. Low serum inorganic phosphorus has been used as an indicator of subclinical trichostrongylosis in lambs, though such a deficiency is not absolutely specific.

3.42

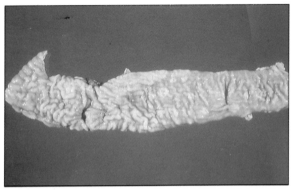

3.42 *Trichostrongylus vitrinus*. Widespread flattening of the duodenal mucosa.

3.4

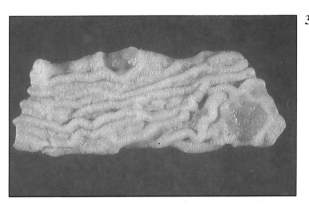

3.43 *Trichostrongylus vitrinus*. 'Fingerprint' lesions of villus atrophy.

Diseases caused by cestodes

3.44

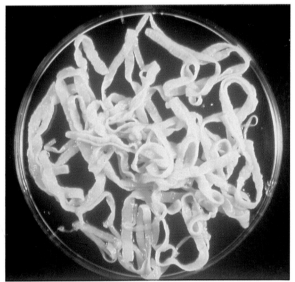

3.44 *Moniezia expansa*. A tapeworm from an infected lamb.

Moniezia expansa and other intestinal cestodes

The common tapeworms that inhabit the intestines of small ruminants are *Moniezia* spp., *Thysaniezia giardi* and *Avitellina* spp. These have oribatid mites or phocids (booklice) as intermediate hosts. They can all cause diarrhoea and ill-thrift in young animals, though mixed infection with nematode worms is more common.

Moniezia expansa is a very large, segmented tapeworm (**3.44, 3.45**), which may be visible through the intestinal wall (**3.46**). Infection is recognised by the passage of the large proglottides in the faeces.

3.45

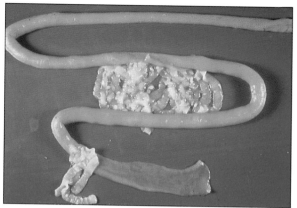

3.45 *Moniezia expansa* **in a lamb intestine.**

3.46

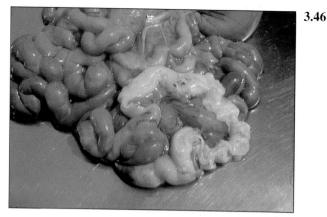

3.46 *Moniezia expansa.* Strobila in the lumen of the intestine.

Thysanosoma actinioides and *Stilesia hepatica*

Thysanosoma actinioides is the fringed tapeworm of North and South America, while *Stilesia hepatica* occurs in Africa and Asia.

The adult forms of these tapeworms occupy the bile ducts (**3.47**), where they become concentrated in sac-like dilatations of the ducts. Generally, they cause no clinical disease, but the biliary fibrosis associated with heavy infections may cause the livers to be condemned in the abattoir.

3.47

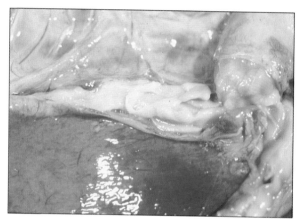

3.47 *Thysanosoma actinioides.* Adult worms in the bile ducts.

Miscellaneous

Limnatis nilotica

Leeches can attach themselves to the oral mucosa of small ruminants (**3.48**), causing quite severe blood loss. The specimen illustrated, *Limnatis nilotica*, is found mainly in the eastern Mediterranean region.

3.48

3.48 *Limnatis nilotica* **in the mouth of a sheep.**

4 Miscellaneous Diseases of the Alimentary System

Ruminants have a complex digestive system compared with most other mammalian species. Diseases of the mouth and teeth interfere with proper rumination, while the complicated anatomy of the forestomachs renders ruminants liable to a wider range of diseases than simple-stomached species. This chapter includes common diseases of the mouth, teeth, pharynx and oesophagus, and then lists conditions affecting the forestomachs, abomasum, intestines and liver, with a short section on the peritoneal cavity.

Conditions affecting the mouth and teeth

Broken mouth

Premature incisor loss, or broken mouth, is an important cause of economic loss in the United Kingdom and Australasia. It can occur at any age from 3–8 years, yet may be totally absent from some flocks. Affected ewes are culled before the end of their useful reproductive life on many farms, particularly on marginal or upland areas.

The disease commences with the development of gingivitis (**4.1**), which worsens soon after incisor eruption. Recession of the gum margins occurs, and the tooth crowns seem very long. Gingivitis proceeds to periodontitis, with progressive deepening of the sulcus around the tooth due to plaque accumulation. This plaque may contain such bacteria as *Bacteroides* and *Fusobacterium* spp. Simultaneously, destruction of collagen in the periodontal ligament supporting the tooth causes loosening and premature loss (**4.2–4.5**).

Multiple factors, which may include the prevalent oral microflora, protein nutrition, host immune responses and trace-element status, probably influence periodontal health and so, along with incisor loss, these may vary from farm to farm.

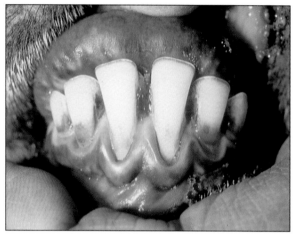

4.1 Broken mouth. Gingivitis in an affected ewe.

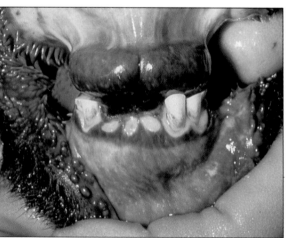

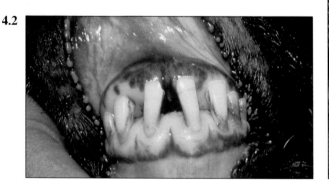

4.2 Broken mouth. Loose teeth.

4.3 Broken mouth. Premature incisor loss.

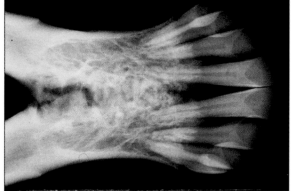

4.4 Broken mouth. Radiograph of a normal jaw.

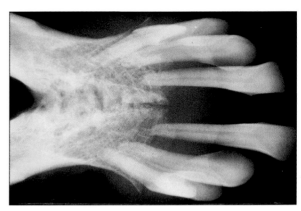

4.5 Broken mouth. Radiograph of loose teeth, showing deficiency of root-supporting structures.

Broken teeth

This disease can occur in animals fed on root crops during their first winter, especially if these become frosted. The disease is a form of caries of the temporary incisors. The brittle incisors break off at the gum margin (**4.6**) and cause feeding problems, with loss of condition. However, eruption of the permanent teeth is seldom affected. Excessive wear of permanent teeth may result from grazing on excessively sandy pastures, in which case affected sheep have to be culled. There is no relationship between this form of excessive wear and broken mouth.

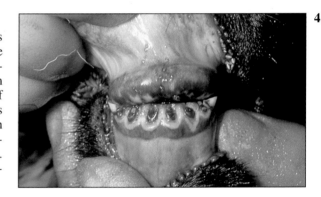

4.6 Caries. Broken temporary teeth in a young sheep.

Caries

Caries of deciduous teeth, with deep pitting of the tooth enamel, is caused by bacterial action, malnutrition or fluorosis. In the latter instance, there is interference with the development of enamel, rather than damage to formed enamel.

Periodontitis involving the molar teeth can cause tooth loss, abscessation or deformity (**4.7, 4.8**), resulting in considerable difficulty in mastication, both in sheep and goats; the resultant unthriftiness leads to early culling.

4.7 Molar tooth wear and decay. Severe lesions in the lower jaw of an aged goat.

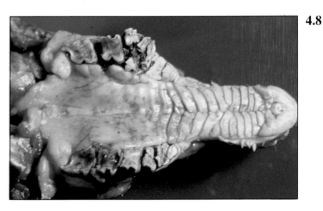

4.8 Molar tooth decay. Similar lesions to **4.7** in the upper cheek teeth of the same goat.

Dental malocclusion

Occlusion is the correct meeting of the incisor teeth and the dental pad. Malocclusion can be hereditary, the result of specific deficiency diseases such as rickets, or may be caused by severe malnutrition. Recovery from malnutrition does not correct the lesion, and defective mastication due to the presence of misshapen teeth (**4.9**) leads to unthriftiness and early culling.

Malocclusion can be part of a complex of dental problems; for example, in the North Island of New Zealand, broken mouth, dentigerous cysts and malocclusion commonly occur together in the same flock.

4.9

4.9 Severe malocclusion of molar teeth.

Dentigerous cysts

Dentigerous cysts are solitary, unilateral cysts that develop in the region of the mandibular incisors of adult sheep (**4.10, 4.11**). The cysts are hard and bony, although the cortical bone may be thinned in 4–5 cm diameter lesions. One or more permanent teeth may be missing and other teeth displaced. Periodontitis or marked incisor wear may also be present.

The cysts are radiolucent, containing milky or caseous foul-smelling material, and sometimes an unerupted tooth. They are lined by stratified squamous epithelium. The aetiology is unknown. They must be differentiated from tumours of the mandible (see Chapter 18).

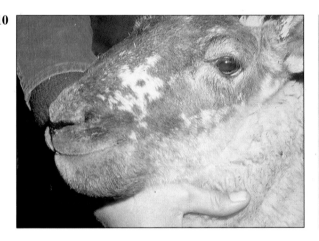

4.10

4.10 Dentigerous cyst. Mandibular swelling in an adult ewe.

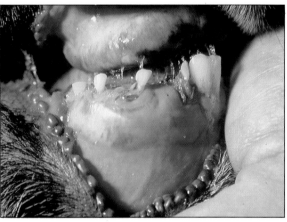

4.11

4.11 Dentigerous cyst. Incisor tooth displacement.

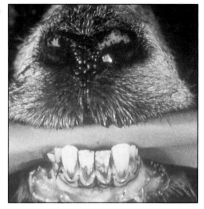

4.12 Fluorosis. Mottling of incisor teeth.

Fluorosis

Industrial pollution by fluorine compounds—for example, from aluminium smelting—can cause interference with normal tooth development. Clinical signs can appear long after the period of exposure to contamination. There is marked pitting of the enamel, with discoloration and chalky mottling (**4.12**). Affected teeth wear much more rapidly than normal. Diagnosis depends on the history of access to contaminated grazings and analysis of herbage and water for fluorine.

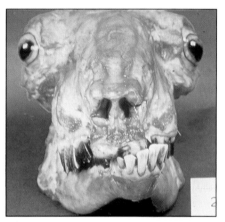

4.13 Mandibular deviation ('shear mouth').

Mandibular deviation

If one of the rami of the mandible is not symmetrical with the other, a serious malocclusion of the molar teeth occurs, resulting in uneven wear and an inability to chew properly.

Sharp edges may form on the buccal aspect of the maxillary molars (**4.13**) and the lingual aspect of the mandibular molars. As this effect becomes more and more accentuated by wear, the teeth tend to pass each other during chewing—'shear mouth'—with loss of cud and progressive unthriftiness.

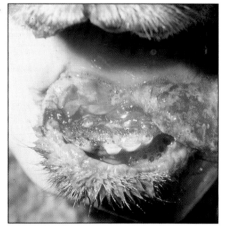

4.14 Orf stomatitis. Oral and lingual erosions.

Orf stomatitis

In some flocks, a particularly virulent strain of the orf virus can cause acute ulceration of the gingivae, undersurface of the tongue or buccal cavity of young lambs or kids (**4.14**). Affected lambs cannot suck, and secondary bacterial infection of the lesions can result in fatalities. (See also Chapter 10).

Salivary mucocele

These are occasionally found in goats, and occur as large fluctuating swellings, usually unilateral (**4.15**) or in the midline of the neck, where they have to be differentiated from caseous lymphadenitis. They result from accumulation of saliva in multiloculated cysts, not by secretion, as the cyst linings lack an epithelium. They may originate from rupture of salivary ducts. Their contents are colourless or blood-tinged and mucinous, and become inspissated with time.

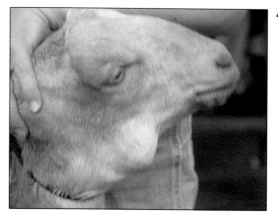

4.15 Salivary mucocele in a young goat.

Conditions affecting the pharyngeal region and oesophagus

Dosing-gun injuries

Careless or inexpert use of dosing guns can cause laceration or even perforation of the pharyngeal region. Lacerations often become infected—for example, with *Fusobacterium necrophorum* or *Actinomyces pyogenes*—causing necrotic ulceration (**4.16**). Perforating wounds are often followed by the development of soft-tissue abscesses (**4.17**), with involvement of the retropharyngeal lymph nodes.

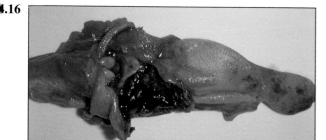

4.16 Pharyngeal necrosis. Dosing-gun injury.

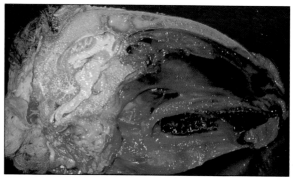

4.17 Pharyngeal abscess.

Choke

Obstruction of the oesophagus by foreign bodies such as plastic material, baling string (**4.18**) or cloth occurs at times in both sheep and goats. Affected animals salivate profusely and make exaggerated swallowing movements. Total obstruction results in accumulation of gas in the rumen and bloat may cause death unless relieved.

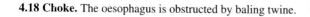

4.18 Choke. The oesophagus is obstructed by baling twine.

4.19

4.19 Choke. The oesophagus is obstructed by sugar-beet pulp.

Feeding of dry sugar-beet pulp can cause death from acute oesophageal obstruction and shock in weaned lambs and adult sheep shortly after feeding. The dry pulp moistened with saliva swells in the oesophagus and totally obstructs it (**4.19**).

4.20

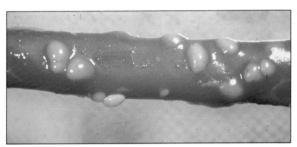

4.20 Mature sarcocysts in a sheep oesophagus.

Sarcosporidiosis

Mature tissue cysts of several *Sarcocystis* spp. are commonly found in the oesophagus of sheep at slaughter (**4.20**). These cause no tissue reaction and have no clinical significance.

Conditions affecting the forestomach

Traumatic reticulitis

Occasionally, the wall of the reticulum is penetrated by a sharp object, such as a piece of wire (**4.21, 4.22**) or sewing needle. Local peritonitis with adhesions can result, and in some instances the foreign body may penetrate the diaphragm to enter the pericardium (see Chapter 13).

4.21

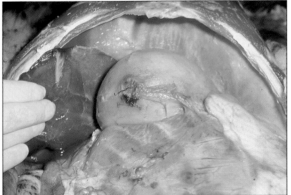

4.21 Traumatic reticulitis. Site of wire penetration.

4.2

4.22 Traumatic reticulitis. Wire exposed by opening the reticulum.

Ruminitis and ruminal ulceration

Severe viral ruminitis with ulceration can occur in bluetongue of sheep and in peste des petits ruminants (see Chapters 2 and 16). The usual cause of ruminitis in non-tropical regions is excessive feeding of rapidly fermentable carbohydrate, usually grain. Sudden introduction of grain feed causes a rapid fall in pH, due to an increase in volatile fatty acids (see Chapter 15). Inflammatory changes in the mucosa develop (**4.23**), and predispose to infections with bacteria, such as *F. necrophorum*, or opportunistic fungi, leading to ulceration and death.

4.23

4.23 Ruminitis in a lamb.

Ruminal impaction

Impaction of the rumen with dry ingesta can occur in both sheep and goats as a manifestation of 'vagus indigestion', or in severe water deprivation. There is total atony of the rumen. Affected animals do not cud, and ruminal sounds are absent due to lack of bacterial fermentation and gas production. Ketonaemia may be present. Impaction by foreign material, such as cloth or plastic, can also occur, but is commoner in goats.

At necropsy, the rumen is distended. The contents are grey and soggy, or dry and fibrous. Foreign material, if present, is easily identified (**4.24**). The carcase is often extremely dehydrated.

4.24

4.24 Ruminal impaction by a plastic bag in a goat.

Ruminal parakeratosis

Feeding of high concentrate rations to sheep during the finishing period can cause ruminal parakeratosis. The lesions probably result from prolonged lowering of the pH, caused by increased concentrations of volatile fatty acids in the rumen contents.

The ruminal papillae are enlarged and hardened (**4.25**), due to the formation of layers of excess keratin, food particles and bacteria, and tend to fuse together in bundles (**4.26**). These may be visible and palpable through the intact rumen wall. The abnormal lining may interfere with the efficiency of feed utilisation.

25

4.25 Ruminal parakeratosis. Generalised enlargement of the ruminal papillae.

4.26

4.26 Ruminal parakeratosis. Close-up of lining, showing fusion of groups of papillae.

4.27

4.27 Woolballs in a sheep rumen.

Ruminal woolballs/hairballs (trichobezoars)

Trichobezoars may be found in animals which have been deprived of dietary fibre or which have pruritis due to ectoparasitism (**4.27**). They are rare in goats, though there is a form in South Africa due to ingestion of a plant, *Chrysocoma* sp., which causes pruritis and excessive shedding of hair. Woolballs seldom cause any clinical symptoms.

Conditions affecting the abomasum

4.28

4.28 Abomasal bloat.

Abomasal bloat

This condition is seen most commonly in watery mouth of young lambs (see Chapter 2), and is associated with hypersecretion of mucin. The mixture of gas and mucin causes marked gurgling on abdominal palpation ('rattle belly'). A chronic form which can be fatal occurs in some orphan lambs at 4–6 weeks of age.

In older animals, gaseous distension of the abomasum (**4.28**) occasionally occurs as a result of pyloric obstruction; for example, after ingestion of the placenta, or by hairballs in goats. It is sometimes seen in forms of indigestion associated with low-fibre diets.

Abomasal clostridial emphysema

This condition, which is fundamentally a gangrenous cellulitis of the abomasal wall (**4.29, 4.30**), is caused

4.29

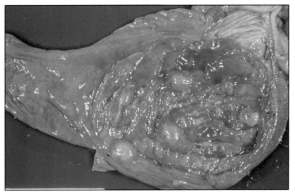

4.29 Abomasal emphysema in a sheep.

4.3

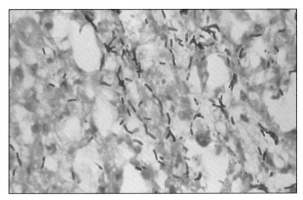

4.30 Abomasal emphysema. Clostridial organisms in the mucosa. Gram stain.

by gas-forming anaerobes other than *Clostridium septicum*. The lesion differs from braxy (see Chapter 2) in that much of the abomasal wall is dissected by gas bullae, in which the vegetative forms of *Clostridium* spp. can be detected.

Abomasal impaction

This condition can occur if poor-quality roughage is fed and water intake restricted, as a sequel to pyloric obstruction—for example, due to foreign bodies—or if abomasal stasis occurs for any reason. It is also seen as one form of an abomasal-emptying defect in Suffolk sheep, in which an hereditary predisposition has been proposed (**4.31**). Reflux of abomasal fluid leads to a diagnostic elevation of ruminal-fluid chloride concentration.

The condition is also recognised in prematurely weaned lambs reared on pelletted rations with a milk replacer. Under these rearing systems, some lambs develop pot bellies, and pine and die after a few weeks. At necropsy, the abomasum is impacted and distended, while the rumen is filled with sour, undigested milk substitute. It has been suggested that oesophageal groove reflex asynchrony is the cause, but this has not been verified.

Abomasal mycosis

This condition can occur as a sequel to chronic loss of acid secretion, or secondary to abomasal ulceration in circumstances where opportunistic fungi can thrive. Antibiotic or steroid therapy may predispose to mycotic infections. The usual organisms involved are zygomycetes (phycomycetes) and include *Mucor, Rhizopus* and *Absidia* spp.

Confluent areas of superficial mucosal necrosis (**4.32**) are found in the abomasum at necropsy. Typical broad, aseptate hyphae can be demonstrated in the tissues by fungal stains (**4.33**).

Abomasal ulceration

Ulcers develop in situations where there is progressive focal necrosis of the abomasal wall. In both sheep and goats, acute ulcers probably are most common in peste des petits ruminants (see Chapter 16). Acute haemorrhagic ulceration also can occur in systemic pasteurellosis (see Chapter 2), and in poisoning with corrosive substances such as arsenic, mercury or zinc.

4.31 Abomasal impaction in a sheep.

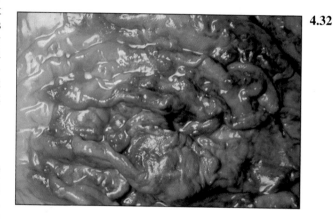

4.32 Mycotic abomasitis. Necrotic mucosa in a sheep abomasum.

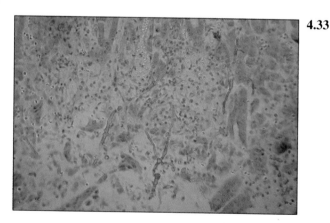

4.33 Mycotic abomasitis. Typical fungal hyphae in the mucosa. Periodic acid–Schiff (PAS).

Chronic ulcers are rare, but are found occasionally as incidental findings at necropsy. They are sometimes complicated by fungal infection (**4.34**).

Necrotic abomasitis

An acute necrotising abomasitis occasionally occurs in young animals following infection with strains of *E. coli* that cause endotoxic damage and thrombosis in mucosal venules (**4.35**). A diffuse abomasitis associated with *Pasteurella haemolytica* A biotype occurs occasionally in unweaned lambs.

4.34

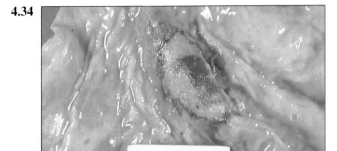

4.35

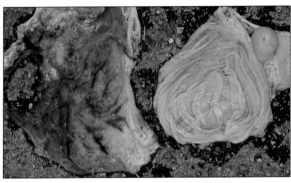

4.34 Mycotic ulcer in a sheep abomasum.

4.35 Necrotic abomasal mucosa (left) in *E. coli* infection.

Infectious diseases that affect the intestine

Campylobacter colitis

Catalase-negative *Campylobacter*-like organisms, tentatively designated *C. ovicolis*, have been associated with weaner colitis in lambs aged 4–12 weeks in Australia. Lambs have a persistent fluid diarrhoea (faecal dry matter < 10%), though they remain bright and alert. Diarrhoea may reach epidemic proportions, and some lambs may develop dependent hypoproteinaemic oedema.

At necropsy, the colon may be inflamed (**4.36**). Microscopic examination of the large intestine shows superficial mucosal erosion, with mononuclear cell infiltrates. Using silver stains or transmission electron microscopy, numerous curved bacteria can be seen adherent to the superficial epithelium (**4.37**). Bacteriological culture of the mucosa in blood agar enriched with mycoplasma agar base or brain–heart infusion agar may result in isolation of *C. ovicolis* or other enteric *Campylobacter* spp.

4.36

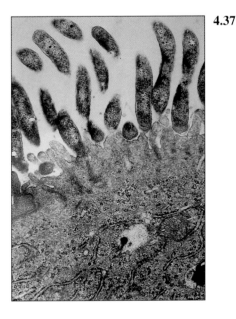

4.37

4.36 Campylobacter (weaner) colitis in a lamb.

4.37 Campylobacter colitis. Transmission electron micrograph showing *Campylobacter* spp. in apposition to the microvillus border.

Caprine herpesviral diarrhoea

A caprine herpesvirus similar to, but antigenically distinct from, bovine herpesvirus-1 can cause severe systemic disease in neonatal kids, with erosions or ulcers in the alimentary tract. Affected kids are febrile and have conjunctivitis, nasal discharge, weakness, abdominal pain, and some may have diarrhoea. Death is common after 1–4 days.

Erosions with a haemorrhagic border may be found in the mouth, pharynx and oesophagus. Longitudinal red erosions occur in the abomasum. Severe lesions of mucosal necrosis can be found in the caecum and colon (**4.38**), often overlaid by a pseudodiphtheritic membrane. On microscopic examination of typical lesions, herpesviral inclusions may be seen in epithelial cells.

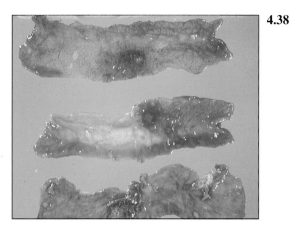

4.38 Caprine herpesviral diarrhoea. Intestinal lesions.

Rotavirus infection

The genus Rotavirus in the family Reoviridae causes diarrhoea in many species, including lambs and kids (**4.39**). The virus infects the jejunal epithelium of animals aged from about 1–4 weeks (**4.40**), causing a transient acute diarrhoea with loss of sucking drive. Viral particles are detectable by routine ELISA or by polyacrylamide gel electrophoresis in the faeces of diarrhoeic lambs, and sometimes older, asymptomatic lambs.

There are no significant macroscopic changes in the gut, apart from the presence of fluid contents. Infected enterocytes are shed from the intestinal villi, which become stunted (**4.41**), with compensatory hypertrophy of the crypts of Lieberkühn. Infiltration by mononuclear cells causes the villi to swell. The normal columnar cells clothing the villi are replaced by cuboidal cells from the crypts.

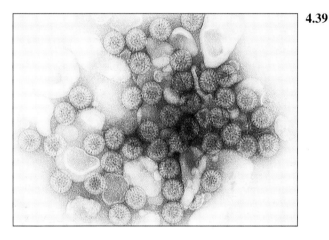

4.39 Rotavirus diarrhoea. Negatively stained viral particles in faeces.

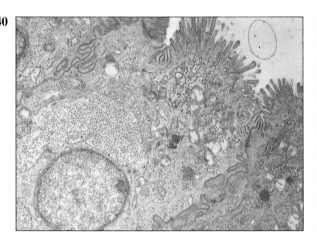

4.40 Rotavirus diarrhoea. Transmission electron micrograph of a virus-infected jejunal enterocyte.

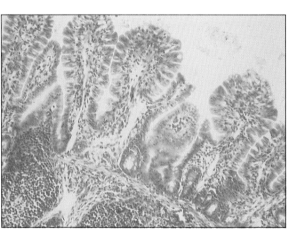

4.41 Rotavirus diarrhoea. Stunted villi clothed with immature epithelial cells. Haematoxylin and eosin.

Miscellaneous intestinal conditions

4.42 Anal stricture in a wether.

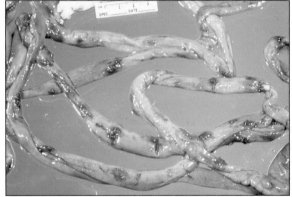

4.43 Buckshot wounds in a goat intestine.

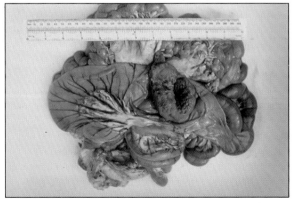

4.44 Intussusception in a lamb.

Anal stricture

This condition can be a sequel to surgical correction of atresia ani. The sheep illustrated had difficulty in defecating, and the constant straining eventually caused stretching and pouching of the perineum (**4.42**).

Buckshot wounds

Goats, and occasionally sheep, can be mistaken for wild game during the hunting season, and may fall victim to gunshot wounds. If pellets perforate the gut (**4.43**), peritonitis can supervene.

Intussusception

In this condition, a segment of bowel telescopes into an outer sheath formed by another, usually distal, segment. A thickened, sausage-shaped structure (**4.44**), up to 20 cm or so in length, may be palpable in lambs or kids. The cause is not always apparent, but irritation due to intestinal parasitism is commonly blamed. Tension on the trapped mesentery causes the inner segment, or intussusceptum, to undergo infarction. Passage of faeces ceases and fluid accumulates in the intestine proximal to the obstruction before death occurs.

Phycomycosis of the gut and mesenteric node

Opportunistic zygomycetes (phycomycetes) occasionally invade the venous circulation of the intestine and colonise the mucosa, causing extensive segmental necrosis. Extension of infection along the afferent lymphatics resulted in infection of the mesenteric lymph nodes (**4.45**) in a sheep with profound leukopenia caused by an experimental tick-borne fever infection.

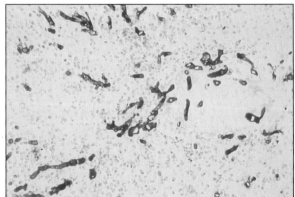

4.45 Phycomycosis of a mesenteric lymph node. Grocott's silver method.

Rectal prolapse

This condition is a fairly common sequel to chronic diarrhoea where there is marked rectal irritation and tenesmus (**4.46**). Coccidiosis is a common underlying cause of rectal prolapse.

Redgut

Grazing on lush pastures—for example, lucerne (alfalfa) in New Zealand or South Africa—is associated with fermentation in the large intestine, leading to intestinal distension and torsion of the anterior mesenteric artery. A large loop of the small intestine becomes acutely congested and necrotic (**4.47**). The duodenum, however, remains unaffected, an important point of differential diagnosis from enterotoxaemia (see Chapter 2). Though the gut contents are dark and bloody, resembling clostridial enterotoxaemia, the condition does not appear to be associated with anaerobic organisms.

Terminal ileitis (regional enteritis)

Terminal ileitis is a sporadic proliferative disease that has been recognised in lambs up to about three months old in the Netherlands, Norway and the United States. The cause is uncertain, though some sources claim to have isolated *Campylobacter* spp. biochemically similar to *C. sputorum* var. *mucosalis* or *C. jejuni* from affected lambs.

Clinical signs include a 'stretching' attitude of the fore- and hindlimbs (**4.48**). Other signs are unthriftiness, intermittent diarrhoea, mild leucocytosis, low total serum proteins, and a low serum albumin/globulin ratio. Necropsy findings include dehydration and emaciation, regional hyperplasia of the ileum (**4.49, 4.50**) with

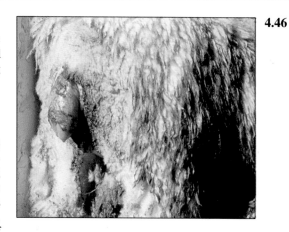

4.46

4.46 Rectal prolapse in a sheep.

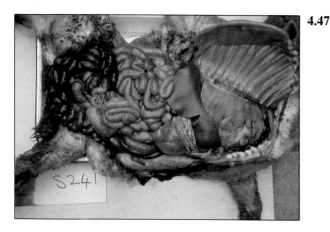

4.47

4.47 Redgut. Note the intense congestion of the affected gut loops.

.48

4.48 Terminal ileitis. 'Stretcher' lamb.

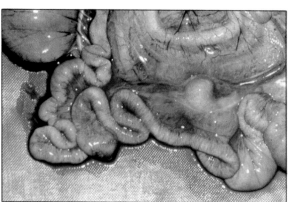

4.49

4.49 Terminal ileitis. Segmental thickening of the ileum.

4.50

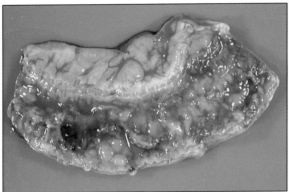

4.51

4.52

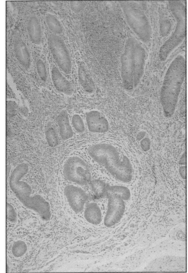

4.50 Terminal ileitis. Mucosal thickening and inflammation.

4.51 Terminal ileitis. Mucosal plaque formation.

4.52 Terminal ileitis. Islets of hypertrophy in affected mucosa. Haematoxylin and eosin.

plaque-like thickenings (**4.51**), and enlargement of the terminal mesenteric lymph node. Some lambs have local mucosal ulceration or diphtheritic inflammation.

Microscopic findings include villus atrophy, crypt hyperplasia with the formation of islands of hypertrophic crypts (**4.52**), hypertrophy of the muscularis and mononuclear cell infiltrates in the lamina propria.

4.53

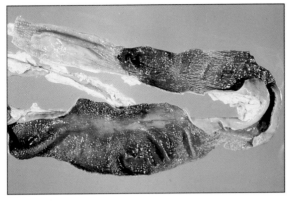

4.53 Uraemic colitis in a sheep.

Uraemic colitis

Colonic damage with severe necrosis occasionally occurs in sheep (**4.53**) with kidney failure. The suggested cause is diffusion from the blood into the colon of urea and other waste products. Ammonia is produced in the intestine by the action of urea-splitting bacteria. The colonic contents smell like urine.

Volvulus

This condition occurs when a loop of intestine becomes twisted to such a degree that blood flow through the local branch of the mesenteric artery is cut off. Affected animals have acute colic (**4.54**) and die rapidly of toxic shock unless the condition is recognised. The small intestine is more commonly involved than the large intestine (**4.55**). A similar condition, redgut (see above), is specifically associated with lush pasture.

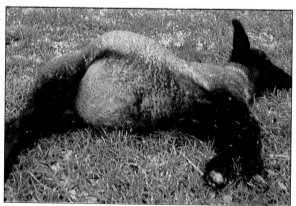

4.54 Volvulus. Lamb recumbent with colic.

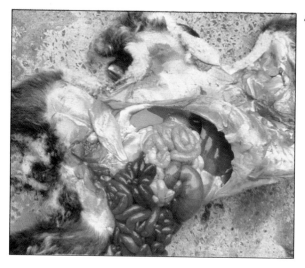

4.55 Volvulus. Typical necropsy appearance.

Conditions that affect the liver and gall bladder

Amyloidosis

This condition is found only occasionally in small ruminants and the cause is often unclear, though it may be a sequel to some chronic debilitating disease. Affected livers may be swollen and pale, and there may also be amyloid deposits in the kidneys (see Chapter 11).

The amyloid is distributed in the portal regions and has a colourless, waxy appearance in routine histological sections. The material is birefringent in polarised light after Congo red staining (**4.56**).

Caseous lymphadenitis

Abscesses can occur occasionally in the liver (**4.57**) as well as in other organs in the visceral form of caseous lymphadenitis (see Chapter 14). These abscesses have to be differentiated from those caused by *Actinomyces pyogenes* and those of melioidosis (see Chapter 11). Sheep and goats affected with the internal abscesses of caseous lymphadenitis are often chronic poor-doers.

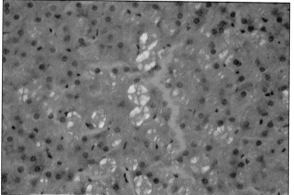

4.56 Amyloidosis of a sheep liver demonstrated by polarised light after Congo red staining.

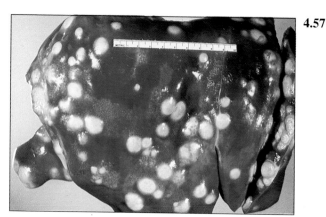

4.57 Caseous lymphadenitis abscesses in the liver.

Cholecystitis

Inflammation of the gall bladder (**4.58**) without associated cholangitis is rare, though it can occur if obstruction occurs due to the pressure of abscesses or tumours in the adjacent liver or hilar lymph node. Fibrinous cholecystitis as a primary lesion also can occur in salmonellosis, as the organism appears to grow well in bile.

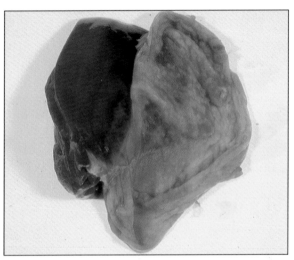

4.58

4.58 Acute cholecystitis.

Coccidia

Endogenous stages of various *Eimeria* species can be found occasionally on histological examination of the liver (**4.59**). These are not known to cause clinical signs at this site.

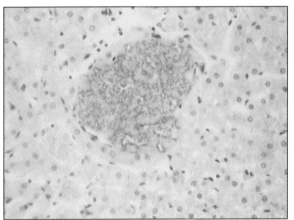

4.59

4.59 Coccidial meront in a sheep liver. Haematoxylin and eosin.

Cysticercus tenuicollis in the liver

Immature metacestode cysts of the dog tapeworm *Taenia hydatigena* migrate through the liver parenchyma, sometimes causing fatal haemorrhage, or encouraging the multiplication of latent *Clostridium novyi* organisms with development of necrotic foci (black disease, see Chapter 2). Many infections are symptomless, and the mature cysts form in the peritoneal cavity overlying the liver capsule (**4.60**) or elsewhere. Eventually the cysts collapse and become calcified (**4.61**).

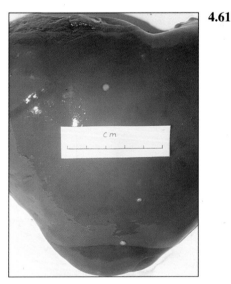

4.60

4.60 Maturing cysts of *Cysticercus tenuicollis* in a sheep liver.

4.61

4.61 Calcified cysts of *Cysticercus tenuicollis* in a sheep liver.

Extramedullary haemopoiesis in young goats

The appearance of the liver in extramedullary haemopoiesis of kids may simulate the lesions of certain forms of phytotoxicity or other poisoning (**4.62**). Histological examination is essential to avoid wrong conclusions in differential diagnosis.

4.62 Extramedullary haemopoiesis in a kid.

Hydatidosis

Sheep and goats can be intermediate hosts of the tapeworm *Echinococcus granulosus*, the definitive host for which is the dog or other canids. Multiple hydatid cysts form in the liver or lungs of the intermediate host (**4.63**), and can cause severe disease. The highly infective cysts are a human health hazard; thus, infected livers are condemned at slaughter.

4.63 Hydatid cysts in a sheep liver.

Liver abscesses

Haematogenous spread of pyogenic organisms following omphalophlebitis or in tick pyaemia is the most common cause of abscesses in the livers of lambs and kids. Omphalogenic abscesses (**4.64, 4.65**) are often restricted to the left lobe.

Surface abscesses can cause local fibrinous peritonitis, with later formation of fibrous adhesions to other abdominal organs. They can also break into the venous system, disseminating infection to the vena cava, heart and lungs. However, in many animals the abscesses become enclosed with fibrous capsules and eventually resolve.

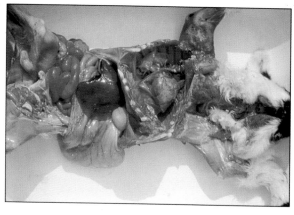

4.64 Abscesses in a lamb liver following omphalophlebitis.

4.65 Abscesses in the liver and retropharyngeal nodes.

4.66

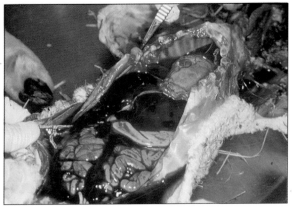

4.66 Ruptured liver with fatal haemorrhage in a newborn lamb.

Liver rupture

Large single lambs are sometimes born dead, with ruptured livers and massive intra-abdominal haemorrhage (**4.66**). Parturition difficulty has been offered as a cause, but there may be a breed predisposition.

4.67

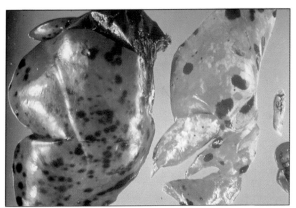

4.67 Melanosis in liver and lungs.

Melanosis

Non-malignant melanosis of the liver and other tissues (see Chapter 7) occasionally occurs in lambs of heavily pigmented breeds. Black areas, several square centimetres in size (**4.67**), are confined to the capsule or stroma. These areas may fade with time.

4.68

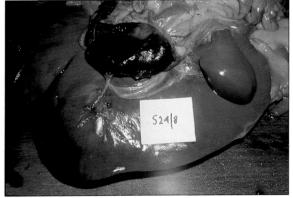

4.68 Torsion of the bile duct.

Torsion of the bile duct

This condition is an occasional finding at necropsy (**4.68**). It is not clear how the gall bladder undergoes torsion, but acute pain and shock are certain to result.

Tyzzer's disease

Focal liver necrosis (**4.69**) caused by *Bacillus piliformis* (Tyzzer's disease) occurs occasionally in kids. The lesions are not pathognomonic, and diagnosis depends on the demonstration of typical elongate gram-negative bacilli at the periphery of the necrotic area (**4.70**).

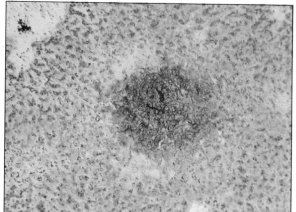

4.69

4.69 Tyzzer's disease in a kid. Liver nodule, Warthin–Starry stain.

4.70

4.70 Tyzzer's disease. *Bacillus piliformis* in a liver lesion. Gram stain.

Conditions that affect the peritoneal cavity

Ascites

Excess fluid in the abdominal cavity is called ascites. It can occur in heart or liver failure, in chronic anaemia due to parasitism, in the protozoal disease heartwater (see Chapter 16), and in certain forms of neoplasia that involve the lymphatic supply to the peritoneum.

Affected animals have distended, tense abdomens (**4.71**), and percussion causes a fluid thrill. When opened, the abdomen is full of pale yellow fluid, which often coagulates soon after exposure to the air. Ascitic fluid must be distinguished from urine in the abdomen caused by rupture of the bladder (see Chapter 11).

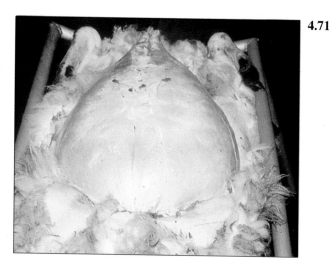

4.71

4.71 Ascites. A sheep abdomen distended with fluid.

4.72

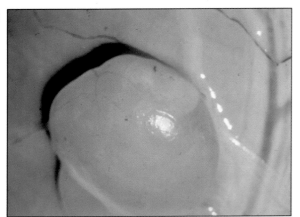

4.72 *Cysticercus tenuicollis.* Cyst attached to sheep intestine.

Tapeworm cysts in the peritoneal cavity

Mature cysts of the dog tapeworm *Taenia hydatigena* can be found in the peritoneal cavity of sheep and goats. These are found most commonly reflected onto the peritoneum overlying the liver (see **4.60**), the diaphragm, or the serous surface of the intestine **(4.72)**.

4.73

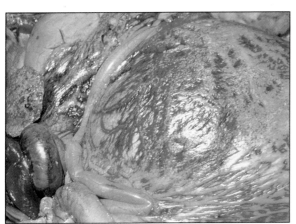

4.73 Acute peritonitis overlying the omasum of a sheep. The aetiology was migration by immature flukes.

Peritonitis

Infection and inflammation of the peritoneum usually occur as a sequel to massive fluke or cestode migration through the liver parenchyma **(4.73)**. Other causes include perforation of abomasal or ruminal ulcers, rupture of the bladder or uterus, and traumatic reticulitis (see above and Chapter 11).

5 Diseases of the Nervous System

Diagnosis of diseases of the central nervous system is hampered to some extent by the fact that many diseases present with similar or identical clinical signs, while, with a few exceptions, macroscopic examination of the brain or spinal cord is seldom of much diagnostic value. An accurate history obviously is of great importance, but in many instances, detailed histopathological examination of particular sites is the only satisfactory means of establishing the cause of death. This chapter attempts to illustrate salient clinical signs associated with a range of infectious, parasitic, metabolic, storage and toxic diseases that affect the nervous system of sheep and goats. Macroscopic pathology is illustrated where appropriate, but in diseases where only microscopic pathology can provide a diagnosis, appropriate photomicrographs are preferred.

Infectious diseases of the central nervous system

Caprine arthritis–encephalitis (CAE)

The CAE retrovirus incorporates itself into the DNA of monocytes in infected goats by means of a reverse transcriptase. Virus expression does not occur until monocytes mature into macrophages. Transmission of infection usually occurs by ingestion of colostrum or milk containing macrophages and thus virus. The virus can cause several different disease syndromes in adult goats, including arthritis (see Chapter 6), progressive interstitial pneumonia (see Chapter 7), and interstitial mastitis (see Chapter 8).

A progressive neurological syndrome is sometimes seen in goat kids, 2 to 4 months old (**5.1**), but has also been recognised in adult goats. Signs may be referable to spinal cord dysfunction (lameness, posterior or tetraparesis, ataxia) or to involvement of the brainstem (dysphagia, circling, torticollis). Protein concentration and leukocyte numbers in the cerebrospinal fluid are increased.

At necropsy, lesions in the spinal cord or brain are usually visible grossly as asymmetrical foci of grey-to-pink discoloration (**5.2, 5.3**). Histologically there is a

5.1

5.2

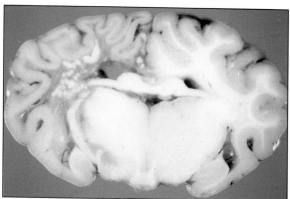

5.1 Caprine arthritis–encephalitis. Young kid with tetraparesis.

5.2 Caprine arthritis–encephalitis. Degeneration of white matter in one-half of the cerebrum and midbrain.

5.3

perivascular infiltration of mononuclear cells accompanied by coagulative necrosis and demyelination of white matter; hence the original name of viral leuko-encephalomyelitis.

5.3 Caprine arthritis–encephalitis. Discoloured lesions in cross-sections of spinal cord.

Listeriosis

The cause of listeric encephalitis is the gram-positive bacterium *Listeria monocytogenes* sensu stricto. This is a facultative anaerobe found in faeces, soil, and in decaying herbage. Infection is more common during the winter than the summer, and inclement weather, intercurrent disease and the feeding of big-bale silage may contribute to the occurrence of disease. The organism can multiply in silage if the pH value is above 5.0 and if excessive amounts of soil are incorporated during preparation.

Infection can cause abortion (see Chapter 9), mastitis, exposure keratitis (see Chapter 12) and septicaemia, but the most frequent manifestation is encephalitis. Circling is the predominant clinical sign. This is generally unidirectional, and an associated head tilt (**5.4**), ear droop with salivation or dysphagia (**5.5**), indicate facial paralysis. Head pressing is another sign (**5.6**). Affected animals soon become recumbent and die. The disease is most common in young adults, but can affect any age from about six weeks old (**5.7**). Infection is believed to gain access through the cranial nerves, particularly the trigeminal, perhaps following infection of the tooth cavity or through buccal injuries.

Histological lesions are most common in the medulla oblongata and in the pons. They are usually unilateral, and consist of areas of malacia and microabscessation (**5.8**) accompanied by substantial lymphocyte/histiocyte cuffs around local blood vessels (**5.9**).

5.4

5.5

5.6

5.4 Listeriosis. Characteristic head tilt in a ewe.

5.5 Listeriosis in a goat. Facial paralysis with inability to masticate.

5.6 Listeriosis. Head pressing in an affected sheep.

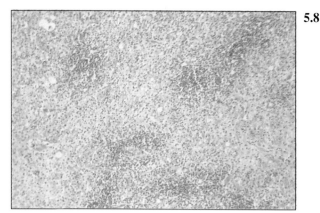

5.7 Listeriosis. Facial paralysis in a lamb.

5.8 Listeriosis. Focal encephalomalacia and microabscessation in the brainstem. Haematoxylin and eosin.

5.9 Listeriosis. Lymphoid cuffing of vasculature in the brain. Haematoxylin and eosin.

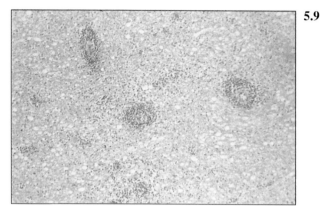

Louping ill

This is a viral disease that affects the central nervous system of many species, including sheep and occasionally goats. The causal agent is one of a group in the family Flaviviridae that causes encephalitides when transmitted by ixodid ticks e.g. *Ixodes ricinus*. All ages are susceptible, but synchrony between lambing-time and the tick rise in many areas means that lambs are probably most vulnerable, though yearlings are also readily infected. Infection may be precipitated by simultaneous transmission by the tick of *Cytoecetes phagocytophila*, the causal agent of tick-borne fever (see Chapter 14), which can cause immunosuppression.

The usual clinical picture after transmission via the bite of an infected tick is fever followed by incoordination, paralysis, convulsions and death within a few days. Some animals have transient ataxia; others survive but with varying degrees of paralysis.

The principal lesion is a widespread non-suppurative meningoencephalomyelitis. Neuron necrosis **(5.10)** and neuronophagia are most commonly found in the motor neurons in the spinal cord, medulla, pons and midbrain and in Purkinje cells in the cerebellar cortex. Focal gliosis with lymphoid perivascular cuffing is widespread in the hindbrain and brainstem. A non-suppurative meningitis is also present in these regions.

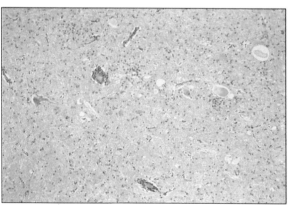

5.10 Louping ill. Non-suppurative meningoencephalitis and neuron necrosis in the brainstem of a sheep. Haematoxylin and eosin.

Rabies

Rabies can occur in both goats and sheep. In Central and South America, the vampire bat, *Desmodus rotundus murinus* (**5.11**), is a reservoir host for the virus. Bats can carry the virus in their salivary glands, and transmit the disease to animals resting at night when they incise the skin to obtain blood. In North America, the raccoon and skunk are important reservoir hosts, while in Europe the sylvatic fox is the main source of infection. Clinically the disease occurs mainly in the paralytic form in sheep and goats.

The nervous tissue lesions are those of a non-suppurative encephalomyelitis. Focal gliosis and perivascular lymphoid cuffing are found in the area from the pons to the hypothalamus, and in the spinal cord. The inclusion bodies of Negri (**5.12**) are characteristic of rabies but do not occur in every case. They are found mainly in the cytoplasm of the Purkinje cells of the cerebellum in ruminants, but may also be seen in the hippocampus or in ganglion cells. Fluorescent antibody tests of brain tissue are also used for rapid diagnosis.

5.11 The vampire bat is an important vector of rabies in Central and South America.

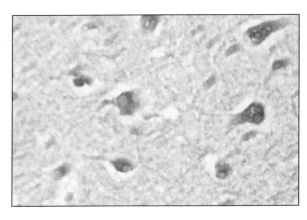

5.12 Rabies. Negri bodies in goat motor neurons. Giemsa stain.

Scrapie

Scrapie can occur independently in sheep and goats, though in the latter it is more commonly found in mixed flocks with sheep. The disease is caused by a self-replicating agent or 'unconventional virus' with remarkable physicochemical stability, which allows it to persist in infected premises. The agent is evidently very small, and, though its genome presumably is nucleic acid, its protein component may be a host product. Host genetic factors determine the course of infection and a scrapie incubation period (Sip) gene has

been shown to control the incubation period in sheep.

The clinical disease is characterised by intense pruritus (**5.13**), with loss of wool on the flanks. Affected sheep tremble and show a nibbling reflex when rubbed on the withers. Incoordination, high-stepping gait and ataxia are also seen, progressing inevitably to recumbency and death.

Striking vacuolation of neurons in the medulla, pons, midbrain and spinal cord is diagnostic of scrapie (**5.14**). Vacuolation is usually bilaterally symmetrical,

5.13 Scrapie. Fleece damaged by rubbing.

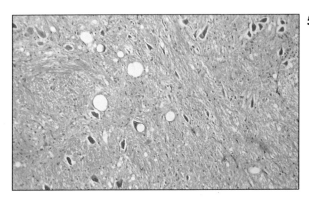

5.14 Scrapie. Typical motor neuron vacuolation in brainstem nucleus. Haematoxylin and eosin.

and is most commonly found in the reticular formation and the medial vestibular, lateral cuneate and papilliform nuclei. Large single vacuoles or multiple vacuoles displace the cell contents peripherally. Spongy change can often be seen in associated grey matter neuropil, and in some sheep cerebrovascular amyloidosis is a prominent feature (**5.15, 5.16**)

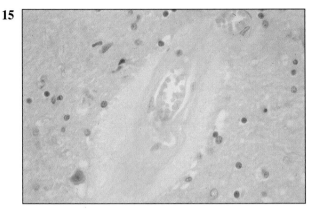

5.15 Scrapie. Perivascular amyloidosis in a sheep brain. Haematoxylin and eosin.

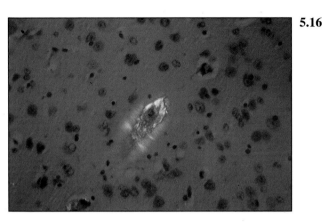

5.16 Scrapie. Birefringence of amyloid plaque after Congo red staining.

Suppurative meningitis

Septicaemic infections by *Streptococcus* spp., *Escherichia coli* and, occasionally, *Pasteurella haemolytica* can cause acute fatal leptomeningitis in lambs and kids. Pyogenic infections can spread from infected navels, while the intestine is the main site for invasive coliforms. In older animals, meningitis can be a complication of wound infection, but quite often is idiopathic.

At necropsy, the meninges are intensely congested (**5.17**), and may have a milky appearance due to swelling and neutrophil infiltration, which is readily confirmed by histological examination (**5.18**).

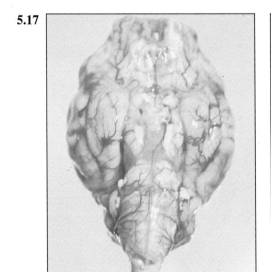

5.17 Acute meningitis in a goat brain.

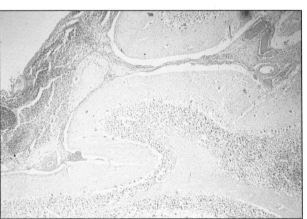

5.18 Acute suppurative meningitis due to streptococcal infection in a kid. Haematoxylin and eosin.

5.19

5.19 Visna. Loss of proprioceptive reflexes has caused knuckling of the fetlock.

Visna

This disease is associated with maedi, a chronic pneumonia of sheep caused by a lentivirus (see Chapter 7). Lesions of visna develop slowly over many months and clinical signs are insidious. Affected sheep show progressive proprioceptive deficit, sometimes with circling or unilateral facial paralysis. Lameness, commencing with knuckling of a fetlock (**5.19**), progresses to incoordination and recumbency.

Changes in the brain and spinal cord are those of non-suppurative meningoencephalitis, with periventricular and perivascular mononuclear cell infiltrates. The choroid plexus also may be infiltrated by mononuclear cells (**5.20**). Glial nodules may be seen in the white matter throughout the brain. Spinal cord lesions are similar but more intense, with myelin breakdown and the formation of cholesterol clefts (**5.21**). In severe cases, there may be focal liquefactive necrosis.

5.20

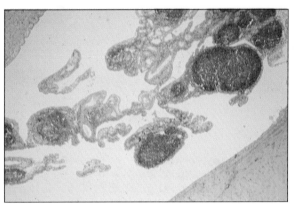

5.20 Visna. Lymphoreticular aggregates in the choroid plexus. Haematoxylin and eosin.

5.

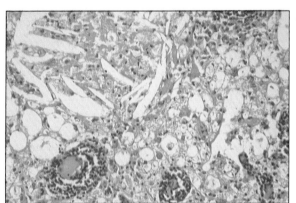

5.21 Visna. Spinal cord lesions include vascular cuffing and residual cholesterol cleft formation after myelin degeneration. Haematoxylin and eosin.

Parasitic diseases of the central nervous system

5.22

5.22 Sheep with sturdy (*Coenurus cerebralis* infection).

Coenuriasis

This disease is caused by *Coenurus cerebralis*, the cystic larval stage of the dog tapeworm *Taenia multiceps*. The acute disease occurs mainly in young lambs, in which symptoms vary between a transient pyrexia and listlessness to severe convulsions and death. The chronic disease in sheep and goats causes progressive unilateral blindness, head tilt with unidirectional circling and incoordination (**5.22**), with recumbency eventually supervening. Softening of the frontal bones may be felt in young animals.

At necropsy, cysts can be seen to have obliterated much of the parenchyma of the cerebral hemisphere (**5.23, 5.24**), with displacement of other regions of the brain. If carried out in reasonable time, surgical removal of the intact cyst will permit recovery.

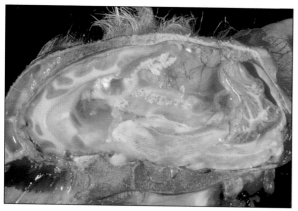

5.23 *Coenurus cerebralis.* Cyst with multiple scolices in a sheep brain.

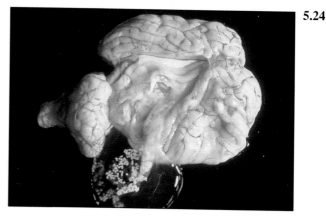

5.24 *Coenurus cerebralis.* The same cyst as in **5.23** partly dissected out.

Oestrus ovis larvae in brain

Larvae of the nasal bot fly, *Oestrus ovis,* can penetrate into the brain if they become trapped while developing in the convolutions of the nasal sinuses of sheep (see Chapter 7). Local necrosis and softening of the affected region, usually the cerebral cortex, are the result of larval presence (**5.25**).

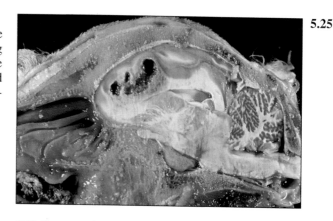

5.25 *Oestrus ovis* **larval migration.** Necrotic brain lesion in a sheep.

Parelaphostrongylus tenuis migration: cerebrospinal nematodiasis

Sheep and goats can become infected with this nematode parasite of the North American white-tailed deer. Snails and slugs on pasture serve as intermediate hosts. Larval migration in aberrant hosts can cause pruritic skin lesions (see Chapter 10), and severe tracking lesions can occur in the spinal cord or brain (**5.26**). The most common clinical presentation is posterior paralysis (**5.27**). The cerebrospinal fluid in such animals often contains numerous eosinophil leukocytes (**5.28**). The diagnosis is confirmed by histological demonstration of the parasite.

Elaphostrongylus rangiferi, a parasite of reindeer, causes a very similar clinical syndrome in sheep and goats. *Setaria digitata* is the cause of cerebrospinal nematodiasis in Southeast Asia.

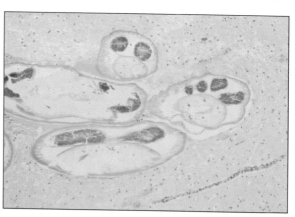

5.26 *Parelaphostrongylus tenuis* **migration.** Parasite cross-section in a goat brain.

5.27

5.28

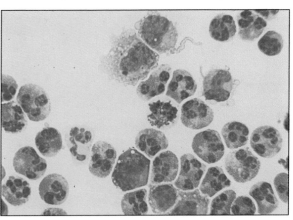

5.27 *Parelaphostrongylus tenuis* **migration.** Posterior paralysis in a sheep.

5.28 *Parelaphostrongylus tenuis* **migration.** Neutrophils and eosinophils in cerebrospinal fluid. Leishman stain.

Sarcocystosis

The definitive hosts for *Sarcocystis* spp. are carnivores which acquire the infection by ingesting the flesh of infected ruminants. This results in the passage of infective oocysts or sporocysts in faeces, where they can contaminate the feed or water. In the sheep, asexual reproductive cycles terminate by invasion of striated muscle cells and nervous tissue, with the formation of tissue cysts.

Infections are usually subclinical, but weakness, ataxia, paralysis and various other signs of central nervous disturbance may be seen (**5.29**), especially if the spinal cord is infected, where lesions may be visible macroscopically (**5.30**).

Microscopically, in the brain or cord, meronts of the parasite can be found (**5.31**), accompanied by a nonsuppurative encephalitis with focal gliosis, haemorrhage and necrosis.

5.29

5.30

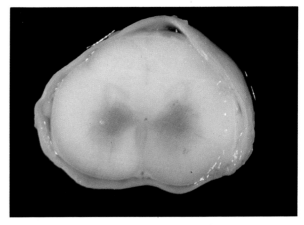

5.29 Sarcocystosis. Paresis caused by infection of the spinal cord.

5.30 Sarcocystosis. Focal lesion in the spinal cord.

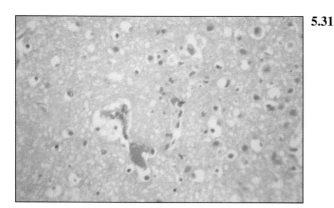

5.31

5.31 Sarcocystosis. Meront of *Sarcocystis* sp. in the brain of an adult sheep. Haematoxylin and eosin.

Toxoplasmosis

Infection of the fetal brain with *Toxoplasma gondii*, a tissue cyst-forming coccidian, causes a characteristic gliosis with focal leucomalacia (**5.32**), which probably represents secondary anoxic damage. Necrotic lesions are most common in cerebral white-matter cores, corpus striatum and cerebellar white matter, and are often mineralised (**5.33**). Toxoplasms are rarely seen in the lesions. Mild meningeal and perivascular infiltrates of lymphoid cells may be seen, and there may be spongy changes in the cerebellar peduncles. A related parasite, *Neosporum caninum*, can cause similar lesions.

Tissue cysts of varying size, containing bradyzoites, may be found in the brains of clinically normal sheep (**5.34**). These represent an asexual, resting phase of the parasite, and evoke little or no reaction in the tissues.

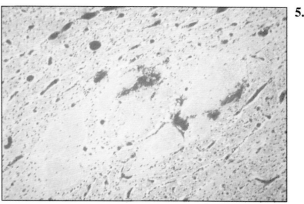

5.32

5.32 Toxoplasmosis. Brain necrosis in the fetus. Haematoxylin and eosin.

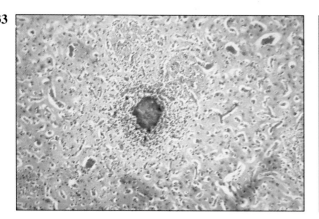

.33

5.33 Toxoplasmosis. Calcified lesion in a fetal lamb brain. Haematoxylin and eosin.

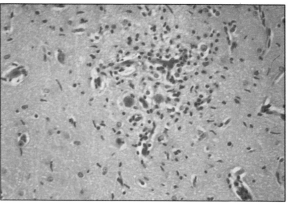

5.34

5.34 Toxoplasmosis. Cyst in the brain of an adult sheep, with minimal glial reaction. Haematoxylin and eosin.

Metabolic diseases that affect the central nervous system

Cerebrocortical necrosis

Cerebrocortical necrosis is a common disease of feedlot and grazing sheep throughout the world; goats are also susceptible. The cause is production in the rumen of thiaminase type 1, as a result of overgrowth of thiaminase-producing bacteria such as *Clostridium sporogenes* or *Bacillus thiaminolyticus*, associated with excess feeding of carbohydrates or grazing on lush pastures. Experimentally, the disease has been induced by feeding sulphur radicles to excess.

Affected animals often are blind and exhibit fine tremors. The head is held high (**5.35**). Later signs include recumbency with convulsions, nystagmus and hypersensitivity to touch or sound. Many sheep are found dead with no premonitory signs, and diagnosis depends on histopathology.

There is marked brain swelling, with laminar necrosis in the cerebral cortex and cerebellar foliae, sometimes obvious at necropsy in coronal sections of the brain (**5.36**). The cerebral hemispheres are swollen and pale with yellow discoloration of some gyri. Where animals have survived for several days, there may be some collapse of the gyral grey matter, resulting in wrinkling of the hemispheres (**5.37**).

When viewed in ultraviolet light, affected areas of cortex often exhibit a white or greenish autofluorescence (**5.38**) that is attributed to accumulation of lipofuscin in the macrophages.

Microscopically, the necrotic laminae have a homogeneous appearance due to neuronal shrinkage. Vacuolation and cavitation of the neuropil is obvious, and the necrotic tissue is sharply demarcated from normal brain tissue (**5.39**).

5.35

5.3●

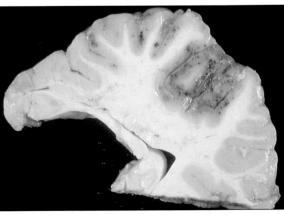

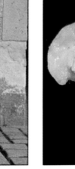

5.37

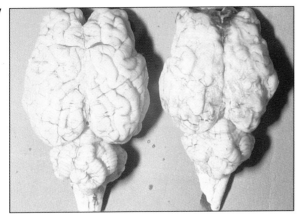

5.35 Cerebrocortical necrosis. A sheep with typical clinical signs.

5.36 Cerebrocortical necrosis. Laminar necrosis of cerebral grey matter.

5.37 Cerebrocortical necrosis. Gyral contraction in long-term disease.

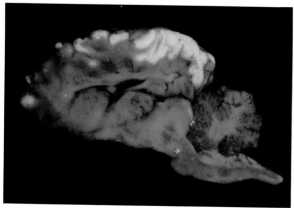

5.38 Cerebrocortical necrosis. Autofluorescence in the cerebral cortex.

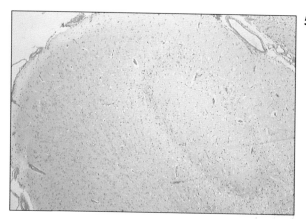

5.39 Cerebrocortical necrosis. Grey-matter necrosis in an affected brain (right half of gyrus). Haematoxylin and eosin.

Swayback

This is an ataxic disorder of newborn lambs and kids caused by low copper status of the dam and progeny (see Chapter 15).

In the congenital form, animals may be stillborn or weak and puny, with nystagmus, facial tremors, depressed pupillary reflexes or grinding of the teeth. Less severely affected animals may be bright and keen to suck, but these become progressively ataxic, with the swaying, stumbling gait that gives the disease its name. Lambs are prone to infections, and may exhibit incomplete extension of the carpal joints.

In the delayed form of swayback in lambs, clinical ataxia may not develop for several months after birth. Also, in these older lambs, a variant, Roberts-type swayback can occur, in which previously robust lambs suddenly develop symptoms of trembling, twitching and aimless wandering, which progress to recumbency, blindness, grinding of the teeth, coma and death.

Cavitation or gelatinous transformation may be seen in the cerebrum of animals with congenital swayback. In the Roberts variant, there is severe cerebral swelling with gyral flattening and herniation of the cerebellar vermis through the foramen magnum. Microscopically, there may be variable hypomyelinogenesis in the brain (**5.40**), and swelling, vacuolation and chromatolysis of large neurons in the red and vestibular nuclei in the brainstem and the reticular formation (**5.41**), the ventral horns in the spinal cord of lambs, and in the cerebellar cortex in kids. In the delayed form, there is hypomyelinogenesis in the spinal cord (**5.42**), with chromatolysis of motor neurons (**5.43**). Altered nerve fibres can be found in brain and spinal cord, with Wallerian-type degeneration.

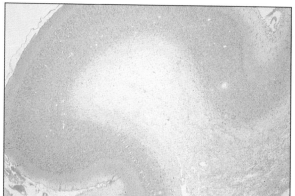

5.40 Congenital swayback. Cerebral hypomyelinogenesis in a newborn lamb. Haematoxylin and eosin.

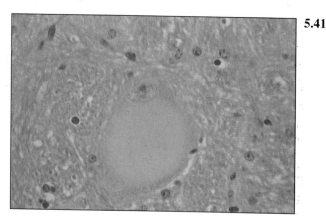

5.41 Congenital swayback. Loss of Nissl substance and peripheral displacement of the nucleus in a motor neuron. Haematoxylin and eosin.

5.42

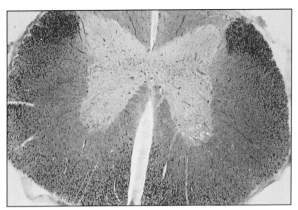

5.4

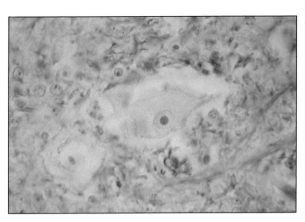

5.42 Delayed swayback. Hypomyelinogenesis in the spinal cord. Marchi stain.

5.43 Delayed swayback. Chromatolysis of ventral horn neurons. Luxol fast blue.

Toxic diseases of the central nervous system

5.44

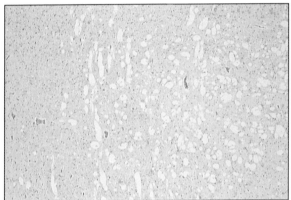

5.44 Copper poisoning. Severe spongy change in cerebral white matter of a sheep. Haematoxylin and eosin.

Copper toxicity

Sheep are susceptible to high dietary copper levels—some breeds, notably the North Ronaldsay, being abnormally so. If excessive intake is prolonged, a haemolytic crisis ensues (see Chapter 17). A proportion of such cases have severe spongiform changes in the white matter of the pons, midbrain and superior cerebellar peduncles (**5.44**).

5.45

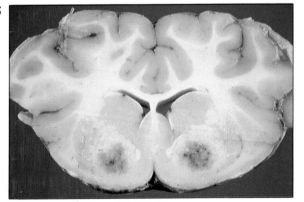

5.45 Focal symmetrical encephalomalacia. Site of malacic foci in a sheep brain.

Focal symmetrical encephalomalacia

This is a bilaterally symmetrical necrotising lesion in the corpus striatum, thalamus, midbrain and cerebellar peduncles of some sheep affected by pulpy kidney disease. The basic lesions are perivascular oedema, vascular haemorrhages and focal malacia, caused by the epsilon toxin of *Clostridium perfringens* type D. These can be seen macroscopically only where haemorrhage is profuse (**5.45**).

The malacic foci are sharply demarcated from normal adjacent brain tissue (**5.46**). The neuropil is vacuolated, and neurons shrunken and degenerate. These changes may be associated with serum protein leakage into the leptomeninges (**5.47**), though this is a nonspecific change.

5.46

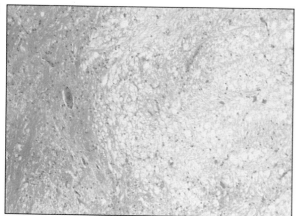

5.47

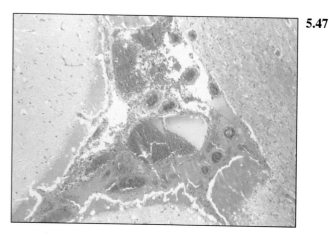

5.46 Focal symmetrical encephalomalacia. Typical malacic changes can be seen on the right. Haematoxylin and eosin.

5.47 Focal symmetrical encephalomalacia. Serum protein leakage into the leptomeninges. Haematoxylin and eosin.

Tetanus

Infection of wounds caused by docking, shearing or castration, etc., with spores of the anaerobic spore-forming bacillus *Clostridium tetani* can result in uptake by local nerve endings of a toxin, tetanospasmin, from bacteria that have become autolysed after germination in the wound. The toxin rapidly travels to and poisons the motor neurons in the ventral horn of the spinal cord.

Stiffness and fine tremors develop within 10 days of injury, and quickly progress to a state of rigidity of the limbs, neck and tail (**5.48**). Animals are hyperaesthetic and constantly go into spasms. Mastication becomes impossible due to spasm of the jaw muscles. Finally, recumbency, opisthotonos and convulsions supervene (**5.49**), and the animal dies of respiratory failure.

5.48

5.49

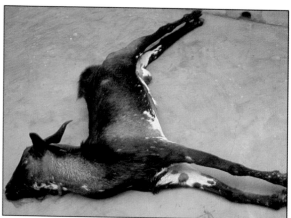

5.48 Tetanus. Note the characteristic tail position.

5.49 Tetanus. Characteristic rigidity in a recumbent buck.

Heritable storage diseases of the central nervous system

β-mannosidosis of Nubian goats

Deficiency of the enzyme β-mannosidase is an important lysosomal disease of Nubian goats, in which it is inherited as an autosomal recessive. This disease is progressively fatal. Ingestion of the toxic plant species *Swainsona, Astragalus* or *Oxytropis* inhibits α-mannosidase in both sheep and goats in a similar manner to the inherited disease of goats (see Chapter 17), except that inhibition can be reversed if exposure to the plants ceases.

Affected goats are unable to rise and have craniofacial abnormalities, including narrowed palpebral fissures, and intention tremors at birth (**5.50**). Cytoplasmic vacuolation is widespread in many tissues, and is particularly obvious in the Purkinje neurons of the cerebellar cortex (**5.51**). There may be white-matter deficiency in the cerebrum and cerebellum, and the ventricles may be dilated.

5.50

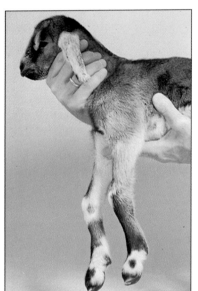

5.51

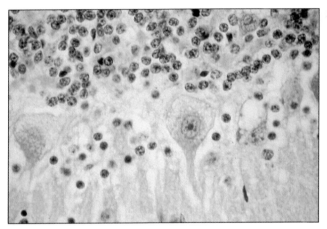

5.50 Goat affected by β-mannosidosis. Note the domed skull and carpal flexion.

5.51 β-mannosidosis in goats, showing cerebellar Purkinje neurons containing storage product. Haematoxylin and eosin.

5.52

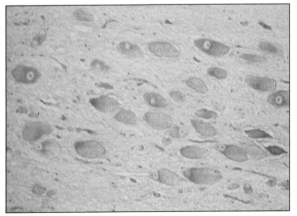

Glycogen storage disease of Corriedales

The precise structure of the enzyme in this heritable enzyme-deficiency disease is uncertain. Affected lambs have locomotor and/or, cardiac disease. Neurons in the brainstem, spinal cord and spinal ganglia are vacuolated (**5.52**), and similar lesions may be found in the heart.

5.52 Glycogen storage disease of Corriedale sheep, showing storage product in the medullary neurons. Haematoxylin and eosin.

Miscellaneous conditions that affect the central nervous system

Cerebral abscess

Cerebral abscesses are caused by the spread of pyogenic bacteria, either as parenteral infections or directly from local trauma.

Infection can also spread from the middle ear, nasal sinuses or other local sites, e.g. following dosing gun injury. Abscesses are space-occupying lesions that reside mainly in the white matter; thus a variety of symptoms may be shown (**5.53**), according to the site of the lesion. Circling, deviation or rotation of the head, drooping of an ear, and unequal pupillary size are possible indications of a brain abscess. Sometimes quite large abscesses can be present with no external signs.

Abscesses often rupture, precipitating terminal illness or obvious symptoms. The presence and site of abscesses are easily confirmed at necropsy (**5.54**).

Disbudding meningitis

Goat kids are routinely disbudded during the first few weeks of life, when the calvarium is thin and there is minimal development of a cornual sinus. Prolonged application of a hot iron or cryosurgery can cause malacia of the underlying cerebral cortex (**5.55**). If bone is killed by heat or application of caustic dehorning paste, bacteria may penetrate the skull directly to the brain and cause abscessation (**5.56**).

5.53 Cerebral abscess. Clinical signs in a lamb.

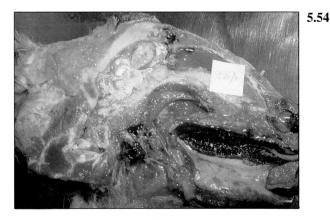

5.54 Cerebral abscess in a lamb brain.

5.55 Post-dehorning malacia. Massive haemorrhage of cerebral cortex induced by excessive heating of the brain. Haematoxylin and eosin.

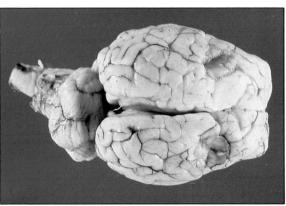

5.56 Post-dehorning abscessation. Malacic areas on each side of the cerebrum underly horn buds destroyed by dehorning paste.

5.57

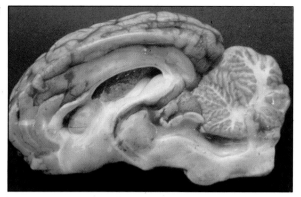

5.57 Abscess in the fourth ventricle of a sheep brain.

5.58

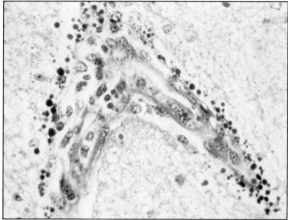

5.58 Heartwater. Periarteriolar oedema and mild lymphoid cuffing. Haematoxylin and eosin.

Fourth ventricular abscesses

Fourth ventricular abscesses are not true abscesses but are the end result of tracking from abscesses in the hypothalamus or cerebrum, which causes a fatal pyoencephaly (**5.57**).

Heartwater oedema

Oedema of the brain, with marked cribriform distension of Virchow–Robin spaces surrounding vascular branches (**5.58**), is a feature of some cases of heartwater, a protozoal disease caused by *Cowdria ruminantium* (see Chapter 16).

Hydranencephaly and porencephaly

Hydranencephaly essentially means total or virtual absence of the cerebral hemispheres, leaving only membranous sacs containing cerebrospinal fluid (**5.59**). Porencephaly is cystic cavitation of the brain, usually the cerebral white matter.

Both of these are generally caused by some destructive process in prenatal life. A variety of arthropod-borne viruses can infect the fetus at a critical point in gestation. These include the causative agents of Akabane disease, bluetongue, Cache Valley virus disease, and Wesselsbron disease (see Chapter 9). Some strains of border disease virus also cause these brain lesions in fetal lambs (**5.60**).

5.59

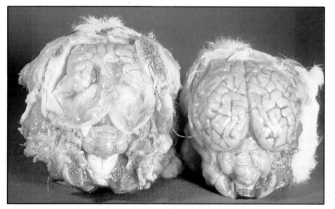

5.59 Hydranencephaly. Virtual replacement of the cerebral hemispheres by membranous sacs (left brain).

5.60

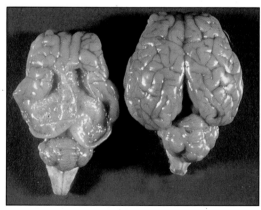

5.60 Hydranencephaly (left) in border disease.

Mesangiocapillary glomerulonephritis brain lesions

Although this disease is an heritable glomerulonephritis (see Chapter 11), affected lambs sometimes have acute lesions in the brain, which cause circling, tremors and convulsions. At necropsy, these can be seen as focal haemorrhages in the median aspects of the cerebral hemispheres (**5.61**).

Microscopically, the lesions are focal haemorrhages with oedema in gyral white-matter cores (**5.62**), possibly an Arthus-type reaction caused by deposition of circulating immune complexes and complement. Oedematous lesions with cellular infiltrates are also found in the choroid plexuses.

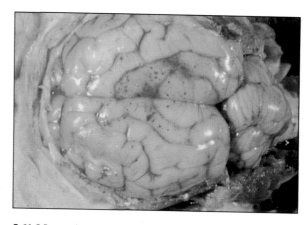

5.61 Mesangiocapillary glomerulonephritis. Brain haemorrhages in a six-week-old Finnish Landrace lamb.

Neonatal meningeal haemorrhage

Meningeal haemorrhages can occur during second-stage labour in oversized lambs or where there is prolonged delivery. Surveys of neonatal mortality in Australia showed that such haemorrhages were frequently found in stillbirths and in anoxic lambs that failed to survive. There may be a breed predisposition; for example, in Dorset sheep.

Typically, haemorrhages and large blood clots can be found in the meninges of the brain and cervical cord (**5.63**) at necropsy of these lambs.

Pituitary abscess

Abscesses in the pituitary arise from infection in the hypophyseal fossa, presumably as a result of trauma, e.g. fighting, or septicaemia. They give rise to vague signs of malaise and nervous disorder but are generally diagnosed only at necropsy (**5.64**).

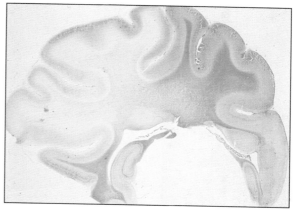

5.62 Mesangiocapillary glomerulonephritis. Gyral white-matter oedema and focal haemorrhage. Haematoxylin and eosin.

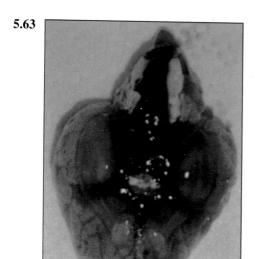

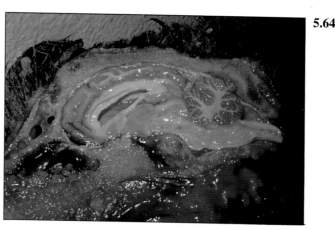

5.63 Neonatal meningeal haemorrhage. Severe basal haemorrhage in a stillborn full-term lamb.

5.64 Pituitary abscess in a sheep brain.

5.65

5.65 Paralysis due to a spinal abscess.

Spinal abscess

These are common in young lambs, especially in outbreaks of tick pyaemia or pyogenic infections from the navel or wounds. In older sheep or goats, abscesses may develop in the spine in disseminated caseous lymphadenitis (see Chapter 14).

As a spinal abscess grows, it impinges on and eventually compresses the spinal cord, causing ataxia and paralysis caudal to the lesion (**5.65**). Thus, animals with lumbar spinal abscesses have posterior paralysis, while those with cervical spinal abscesses are quadriplegic. The site of a spinal abscess is usually readily determined at necropsy (**5.66, 5.67**) and is quite often situated above the heart or kidney. Small abscesses may be quite difficult to detect.

5.66

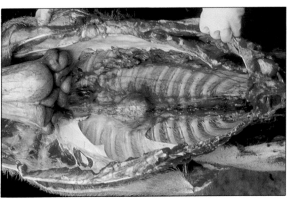

5.66 Position of a spinal abscess at necropsy.

5.67

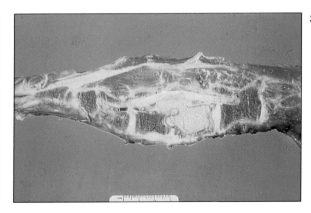

5.67 Spinal abscess in caseous lymphadenitis.

5.68

5.68 Spinal fracture in a lamb.

Spinal fracture

Most spinal fractures are the result of trauma, but nutritional factors, such as protein and/or copper deficiency, mineral imbalance, prolonged poisoning by lead salts or chronic subclinical parasitism, may contribute to an undue porotic state of the bones, thus predisposing to fracture. Cervical fracture can occur during fighting amongst rams, while lumbar fractures are commonest in lambs (**5.68**).

Clinically-affected animals are usually paralysed caudal to the fracture site, due to pressure by broken bones with associated haemorrhage into the spinal canal.

6 Diseases of the Locomotor System

Diseases of bone, apart from those caused by injury, tend to result from complex interactions of mineral imbalances and vitamins, and may be influenced by parasitism. In contrast, diseases of joints tend to be the result of infection, especially in young animals. Painful joint conditions can create serious welfare problems. Some hereditary bone conditions are dealt with in Chapter 1, and congenital arthrogryposis in Chapter 9. Bone tumours are covered in Chapter 18. Several parasitic infections which affect muscle are illustrated in this chapter, but we have preferred to present nutritional myopathy in Chapter 15.

Diseases of bones

Rickets

Rickets is a metabolic disease of growing bone. Newly formed bone matrix in the epiphyses of long bones fails to mineralise (6.1), due to deficiency of calcium, phosphorus or vitamin D. Growth plates become thickened with a build-up of unmineralised cartilage, and there is increased space between bone spicules (6.2); consequently, bones are soft. Marked enlargement of the costochondral junctions (rachitic rosary) is occasionally seen, but commonly the rib lesion is a more subtle white line, demonstrated by incising the distal end of a rib longitudinally (6.3).

Affected lambs and kids may show stiffness or shifting lameness. Pathological fractures of long bones or vertebrae are common.

6.1

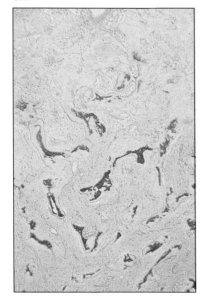

6.1 Rickets. Unstained osteoid borders surround black trabecular bone and bone growth is disorganised. Von Kossa stain.

6.2

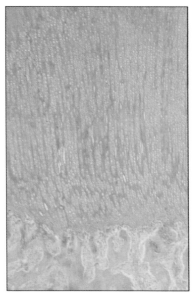

6.2 Rickets. Irregular growth plate and increased space between bone spicules. Cartilage is normal but increased in depth. Haematoxylin and eosin.

6.3

6.3 Osteodystrophic line. Numerous mineral imbalances produce degenerate bone formation at or above the costochondral junction.

6.4   **6.5**

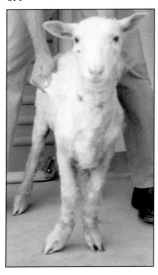

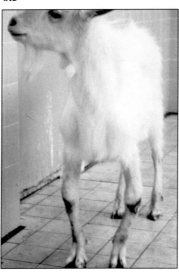

Bentleg, bowie

Growing lambs and kids with osteodystrophic disease often develop distortion of the limbs, loosely termed bentleg or bowie (**6.4, 6.5**). Both genu varum and genu valgum (knock-kneed) forms occur. Hindlimbs are less commonly affected. The aetiology is unclear and probably variable, but a phosphorus deficiency may be involved in suckling lambs on unimproved pasture.

Similar deformation of the limbs occurs in feeder lambs that are gaining weight very rapidly and in yearling goats late pregnant with multiple fetuses. Presumably there is too much weight being applied to an immature skeleton under these circumstances.

6.4 Bentleg. Lamb demonstrating inward bending of the limbs at the fetlock.

6.5 Bentleg. Goat with marked outward bowing of the forelimbs at the carpus.

Osteoporosis

Osteoporosis, or a reduced amount of normally mineralised bone matrix, can occur in growing and in adult animals. Trabecular bone is reduced and cortical bone is more porous than usual (**6.6, 6.7**). These changes increase the risk of fractures. One form of osteoporosis is cappie, or double scalp, in which there is a softening of the frontal bone; this condition is seen in sheep in Scotland. The aetiology is unclear, but dietary mineral imbalances and vitamin D deficiency in northern latitudes are presumably involved. Osteoporosis with brittleness of bones and a tendency for spontaneous fracture is a feature of chronic copper deficiency and lead toxicity (see Chapters 15 and 17). Deficiency of bone matrix occurs in growing lambs subclinically infected with nematode parasites, probably due to a combination of endogenous protein loss caused by the direct action of the worms, with reduced digestibility of dietary crude protein.

6.6

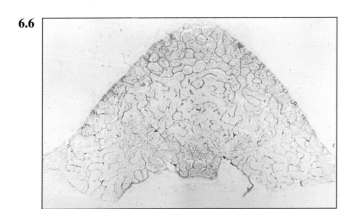

6.7

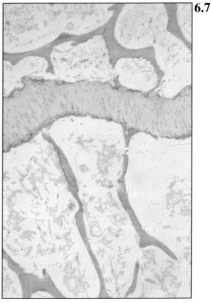

6.6 Osteoporosis. Reduction in trabecular bone in a lumbar vertebra..

6.7 Osteoporosis. Bone spicules are decreased in number and thin but normally mineralised. Note the premature formation of a bony plate beneath the epiphyseal cartilage. Haematoxylin and eosin.

Spondylosis

Extensive new bone formation and ventral
bridging between vertebrae occasionally
occur (6.8). In some instances, it may be
associated with excessive mineral intake by
mature male animals. The spinal column
becomes abnormally rigid and therefore
subject to fracture.

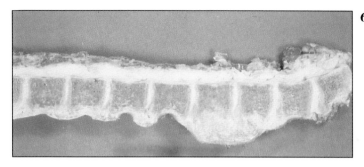

6.8

6.8 Spondylosis. Longitudinal section of a spinal column, demonstrating
ventral ankylosis between vertebrae.

Skull and neck fractures

Animals with osteodystrophic diseases and those subject to skull trauma (such as fighting rams) may suffer
fractures to the base of the skull (6.9, 6.10) or neck.
Many animals with broken necks are found dead.

Signs of acute paralysis might be expected in the live
animal, but basilar fractures have also been proposed
to lead to pituitary abscessation (see Chapter 5).

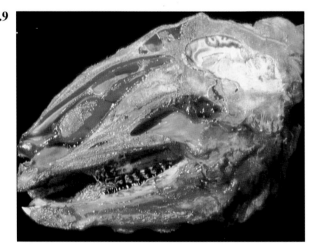

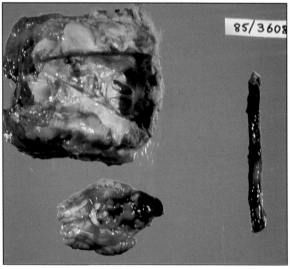

6.9 Skull fracture. A fracture line is visible in front of the
pituitary fossa, which contains an abscess.

6.10 Skull and neck fracture. Massive haemorrhage around
the spinal cord and base of the brain.

Diseases of joints and tendons

Arthritis

Arthritis in sheep and goats is usually of infectious origin. Single joints may be involved, as from penetrating wounds, but typically the infection is blood-borne and a polyarthritis occurs. Joint fluid is an excellent growth medium for bacteria. Occasionally, trauma to a joint such as the stifle causes ligament damage and joint instability followed by degenerative arthritis. Some old sheep and goats show progressive pain and stiffness due to degenerative, age-related arthritis involving multiple joints.

Infectious arthritis is usually accompanied by joint swelling (**6.11**). The synovial membrane lining the joint has a villous structure and often proliferates with infection. Inflammatory products within the joint contribute to erosion of cartilage (**6.12**). In chronic infections, the articular surface may be covered by granulation tissue (pannus, **6.13**) and surrounded by osteophytic new bone growth (**6.14**). An osteochondritis dissecans lesion would appear similar.

6.11

6.11 Bacterial arthritis. The right carpal joint is distended due to infection with *Actinomyces pyogenes*.

6.12

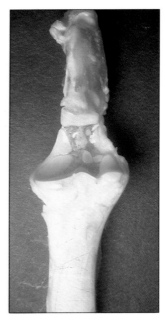

6.12 Arthritis at the elbow joint. Synovial hyperplasia with slight erosion of the articular cartilage.

6.13

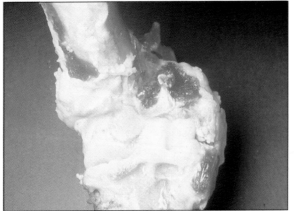

6.13 Arthritis at the elbow joint. A plaque-like pannus lesion is present on one condyle.

6.

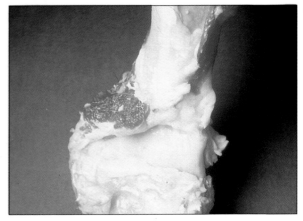

6.14 Chronic osteoarthritis at the elbow joint. The joint is ankylosed and surrounded by nodular masses of osteophytes.

Joint-ill

Joint-ill (**6.15**) is a suppurative polyarthritis of young lambs and kids that results from infection of the navel by pathogenic bacteria in the environment. Sudden onset of lameness can occur in many lambs simultaneously. These animals are usually depressed and febrile. The joints most commonly affected—the fetlocks, carpi, hocks (**6.16**) and stifles—are hot and painful, with fluctuating swellings within distended joint capsules. Affected animals may die, or at best are left unthrifty and stiff.

When incised (**6.17**), affected joints ooze greenish or greyish pus. Suppurative foci may also be found in the lungs, kidneys, myocardium or meninges, and some lambs develop vegetations on the heart valves. The organisms most commonly isolated are *Actinomyces pyogenes* and haemolytic streptococci.

Tick pyaemia (cripples) is a similar condition in which pyogenic bacteria, especially *Staphylococcus aureus*, may be introduced to the bloodstream of young lambs by the bite of the nymphal stage of the tick *Ixodes ricinus*. As in joint-ill, suppurative polyarthritis is the main sequel, with abscesses forming in various other organs, including the spinal column (see **2.41**). However, compared with joint-ill, the disease can be found in lambs up to three months old, and only on pastures or hills where the tick is active.

6.15

6.15 Joint-ill in a young lamb. Carpal and hock joints are distended and the lamb has difficulty standing.

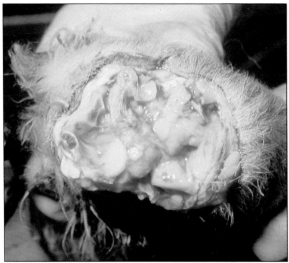

6.17

6.17 Joint-ill in a lamb. Synovial proliferation with chronic infection. Some cartilage erosion is present but hard to distinguish from normal synovial fossae in the joint.

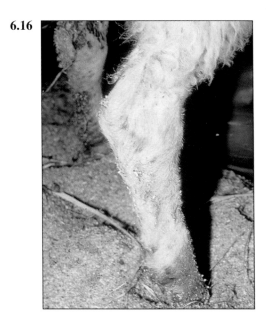

6.16

6.16 Joint-ill in a lamb. Distension of the anterior aspect of the hock joint.

Erysipelothrix arthritis

Infection of skin wounds by soil-borne *Erysipelothrix rhusiopathiae* can cause severe arthritis in lambs and sheep up to about six months of age. Outbreaks may follow procedures such as docking, shearing or dipping. In the latter, the bad practice of re-using stagnant dip from a previous dipping is an important predisposing factor. Young lambs can acquire infection through the navel.

Affected animals are lame or stiff (**6.18**), and may have a high fever. In acute infections the joints are painful but not swollen, though they may contain a greenish turbid fluid (**6.19**), but in neglected cases the joints are enlarged and ankylosed.

Incision of chronically infected joints reveals a greenish, semi-coagulated fibrinous exudate, accompanied by fibrosis and osteophyte formation. Affected sheep are unthrifty.

6.18

6.18 *Erysipelothrix* arthritis. Distended carpal joints, lowered head and painful attitude as the lamb attempts to walk on its toes.

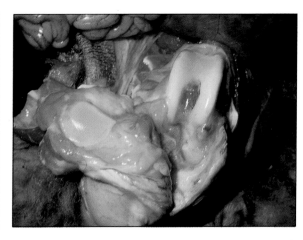

6.19 *Erysipelothrix* arthritis. Non-purulent gelatinous exudate in the stifle joint and synovial proliferation around the patella.

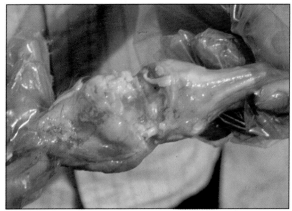

6.20

6.20 Mycoplasma arthritis. Acute fibrinous arthritis in a goat, caused by *Mycoplasma agalactiae*.

Mycoplasma and chlamydial polyarthritis

Several species of mycoplasma, including *Mycoplasma mycoides* subsp. *mycoides* large colony type, *M. capricolum*, *M. putrifaciens,* and *M. agalactiae*, cause septicaemia and polyarthritis, especially in goat kids (**6.20**). Many animals may be affected simultaneously, especially if fed pooled raw milk originating from carrier does. Arthritis may be accompanied by pleuropneumonia or keratoconjunctivitis, induced by the same mycoplasmas (see Chapters 7 and 8). Affected animals are typically febrile. The joint fluid is fibrinopurulent. Culture for confirmation of the diagnosis is most successful when performed early in the course of infection, using appropriate mycoplasma media.

In feedlot lambs, a similar polyarthritis often accompanied by conjunctivitis (see Chapter 12) is caused by immunotype-2 strains of *Chlamydia psittaci*, which are different from immunotype-1 abortion strains. Lambs are lame or stiff and stand humped up or remain recumbent for long periods. Joint fluid contains fibrin and increased mononuclear cells. Diagnosis is by fluorescent antibody test or culture in embryonated eggs or tissue-culture systems.

Caprine arthritis–encephalitis (CAE viral infection)

The CAE retrovirus has a predilection for joints but also causes lesions in nervous tissue, udders and lungs. Transmission commonly results from consumption of colostrum or milk containing infected macrophages. Many goats with CAE infection remain asymptomatic or exhibit only carpal hygromas without lameness. Clinical signs of joint involvement rarely develop before one year of age.

Initially there is soft-tissue swelling (6.21) with proliferation of synovial membrane. Many free-floating concretions (rice bodies) accumulate in joints and bursae (6.22). In animals that become lame, signs may develop suddenly or very gradually. In chronically involved joints, extensive osteophyte formation and mineralisation of periarticular tissues occur (6.23, 6.24). These animals become debilitated and often walk on their carpi, with resulting tendon contractures and hoof overgrowth. Although less commonly recognised, the maedi-visna (ovine progressive pneumonia) virus can cause similar joint lesions (6.25).

A symmetrical distension of the atlantal bursa (6.26), beneath the nuchal ligament, is an uncommon but distinctive clinical finding in goats with CAE. Floaters within the bursa may undergo mineralisation.

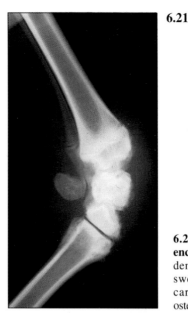

6.21

6.21 Caprine arthritis-encephalitis. Radiograph demonstrates soft-tissue swelling anterior to the carpal joint and early osteophyte formation.

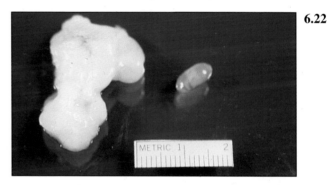

6.22

6.22 Caprine arthritis-encephalitis. Free-floating concretion and portion of hypertrophied synovial membrane from the carpus of a mature goat.

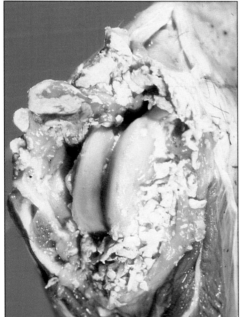

6.23

6.23 Caprine arthritis-encephalitis. Opened stifle joint with marked proliferation of the synovial membrane and numerous rice bodies.

6.24 Caprine arthritis-encephalitis. Advanced lesion with mineralisation of the joint capsule and tendon sheaths.

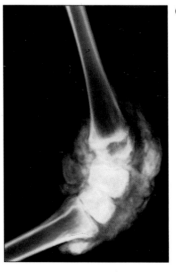

6.24

6.25

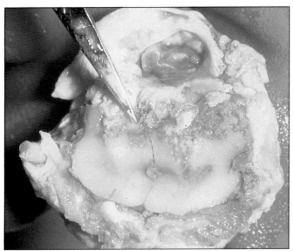

6.2

6.25 Maedi-visna arthritis. Sheep carpal joint with marked synovial hypertrophy and mineralisation.

6.26 Caprine arthritis-encephalitis. Bilateral distension of the atlantal bursa.

Suppurative tenosynovitis

In some animals with septicaemia or penetrating wounds, infection localises in the synovial lining of tendon sheaths rather than in joints (**6.27, 6.28**). Differentiation from arthritis is aided by the more localised, longitudinally oriented nature of the swelling in the live animal and the absence of joint lesions at necropsy.

6.27

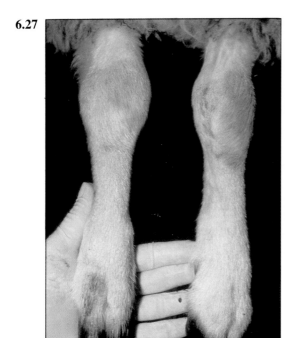

6.2

6.27 Suppurative tenosynovitis. Swelling is localised anterior to the lamb's carpal joints.

6.28 Suppurative tenosynovitis. The opened carpal joint reveals normal articular surfaces but thickening of the extensor tendon sheaths over the carpus.

Contracted tendons

Lambs and kids are occasionally born with contracture of the forelimbs involving the flexor tendons and sometimes the carpal joint capsule (**6.29**). The aetiology is poorly understood, but in some individuals large size apparently restricts normal limb movement *in utero*. Joints move but cannot be extended fully without application of much force. Arthrogryposis associated with fetal brain lesions (see Chapter 9) is a separate condition where the joints are fused in a flexed position.

Tendon contracture also occurs in the forelimbs when sheep or goats carry a limb without weight-bearing, because of a painful condition, or remain recumbent for several days.

6.29 Contracted tendons. This kid was unable to fully extend the forelimbs from birth.

Diseases of muscle

Blackleg

Infection of the limb muscles by the anaerobe *Clostridium chauvoei* appears to be rare in sheep and goats. There is blackening of the affected muscles (**6.30**) and a characteristic rancid odour. Gas production in the tissues is a feature of rapid autolysis in these species regardless of the cause of death.

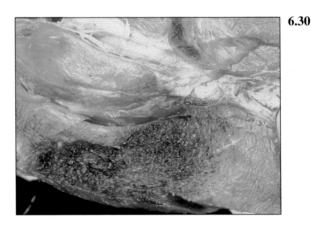

6.30 Blackleg. Note the blackened area of acute muscle necrosis.

Cysticercosis, sheep measles

Larvae of *Taenia ovis*, a tapeworm of dogs and foxes, can invade skeletal or cardiac muscle of sheep (**6.31**). The cysticerci, or bladder worms, measure 10 mm × 20 mm and contain a single invaginated scolex and clear fluid. The parasites are usually an incidental finding at meat inspection or necropsy of young animals. The life cycle is completed when the cysticerci develop into tapeworms in the dog that ingests uncooked meat from a sheep with measles.

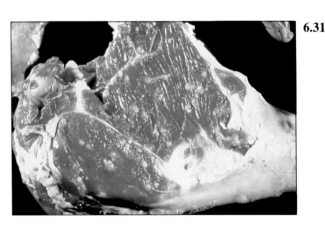

6.31 Cysticercosis of muscle. *Taenia ovis* cysticerci are visible as pale nodules in the muscle of this lamb from New Zealand.

6.32

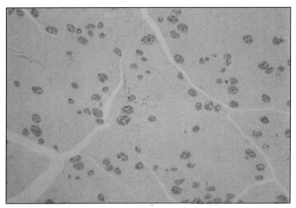

6.32 Sarcocystosis. The oval basophilic parasitic structures are incidental findings in a sheep myocardium. Haematoxylin and eosin.

Sarcocystosis

The sheep is the intermediate host of the tissue cyst-forming coccidial species *Sarcocystis tenella, S. arieti-canis, S. gigantea* and *S. medusiformis,* and mature cysts of these species can be found in ovine muscle cells. These are formed when merozoites from infected endothelial cells invade the fibres during asexual reproduction of the parasite **(6.32)**. *Sarcocystis capri-canis* cysts have been identified in goat muscle **(6.33)**.

The skeletal muscles, tongue, oesophagus and myocardium often contain large numbers of these cysts, each packed with cystozoites. The cysts of the macrocyst species *S. gigantea* and *S. medusiformis* may be visible macroscopically as white streaks or spots in the muscles **(6.34)**.

6.33

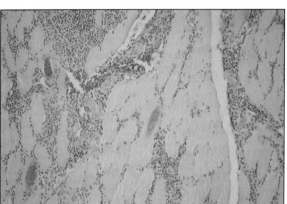

6.33 Sarcocystosis. Severe myositis with several clumps of parasites and marked cellular response in an experimentally infected goat. Haematoxylin and eosin.

6.34

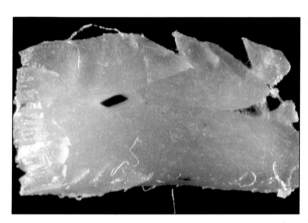

6.34 Sarcocystosis. Faint white streaks are visible in hind limb muscles.

6.35

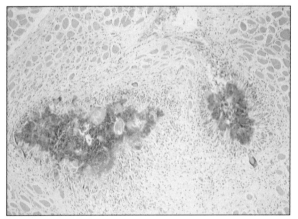

6.35 Eosinophilic myositis. Pyogranulomatous inflammation with giant cells surrounds remnants of sarcocysts in a sheep skeletal muscle. Haematoxylin and eosin.

Eosinophilic myositis

If cysticerci or sarcocysts die in muscle, the resulting inflammatory lesions **(6.35)** are often green. These are termed eosinophilic myositis, especially by meat inspectors.

Umbilical hernia and umbilical abscess

The muscles of the ventral abdominal wall sometimes fail to close at the umbilicus, either because of an hereditary defect or because structures making up the umbilical cord were previously swollen due to neonatal omphalophlebitis. An uncomplicated hernia is usually soft, fluctuant and painless (**6.36**). The hernia contents (often omentum or abomasum) may be manually reducible into the abdomen. An umbilical abscess causes a swelling in the same location, but the mass is usually not reducible. Deep abdominal palpation may identify thickening of umbilical vessel remnants leading forwards to the liver or posteriorly towards the bladder. Aspiration may yield pus.

6.36

6.36 Umbilical hernia. This hernia was soft and reducible.

Flank hernia

Traumatic ventral hernias, which commonly lead to dystocia, are usually caudal to the umbilicus and in a flank or paramedian location (**6.37**, see also **8.35**).

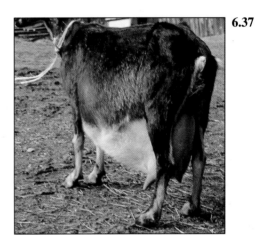

6.37

6.37 Abdominal hernia. The ventral abdomen just anterior to the goat's udder hangs abnormally low.

Ruptured gastrocnemius muscle

Traumatic rupture of fibres in the gastrocnemius muscle occurs with a sudden, extremely forceful attempt to jump. The condition is easily diagnosed because the hocks rest on the ground when the animal stands (**6.38**).

6.38

6.38 Ruptured gastrocnemius muscle. The hocks of this parturient doe are on the ground because of bilateral muscle rupture.

Diseases of the foot

Footrot is one of the most important and prevalent infectious causes of lameness in sheep and to a lesser extent in goats. It is discussed fully in Chapter 10 (see **10.10–10.15**) with other skin diseases affecting the coronary band region, such as orf (see **10.2–10.5**) and dermatophilosis (see **10.22–10.26**). Foot-and-mouth disease and bluetongue (see Chapter 2) may also be accompanied by lameness.

Foot abscess

Penetrating wounds, excessive trimming, or severe footrot may permit bacteria such as *Fusobacterium necrophorum* and *Actinomyces pyogenes* to reach deep structures in the hoof. However, in many instances no obvious cause can be identified. Inflammation and accumulation of pus within the hoof, distal phalangeal joint, or tendons cause pain and the animal tends to hold the affected foot off the ground. Grazing is restricted. Abscesses involving the joint or the sensitive laminae of the wall commonly break and drain at the coronary band (**6.39**).

6.3¶

6.39 Foot abscess. Pus is draining at the coronary band.

Laminitis

Overfeeding of highly fermentable carbohydrate rations may cause acidosis, one of the sequelae of which is laminitis. Ewes fattening on grain, lambs on protein-supplemented carbohydrates, goats being fed for high milk production or rams being fed for show are at risk.

Affected animals are very lame and prefer to lie down or walk on their carpi (**6.40**). The feet are hot and painful in the acute stage, and the animal may be greatly distressed. Neglect will lead to permanent hoof deformities (**6.41**).

 6.40

6.40 Laminitis. This goat prefers to walk on its knees.

6.41

6.41 Laminitis. Separation of hoof wall from the sensitive laminae, with the resulting cavity opened by corrective trimming. This ewe was transferred from dry lot to lush grass six months previously.

7 Diseases of the Respiratory System

The majority of respiratory diseases are the result of infections: viral, bacterial, fungal or parasitic. Many are associated with chronic unthriftiness, while others cause high mortality, often of epidemic proportions. Additionally, there are numerous diseases of other systems in which secondary involvement of the respiratory system causes lesions which can be recognised at necropsy. This chapter includes a number of the most common respiratory infections of sheep and goats, with a selection of less common pulmonary diseases.

Acute viral diseases of the respiratory system

Adenoviral pneumonia

Six serotypes of ovine adenovirus and several untyped isolates have been identified. Several adenoviral serotypes have been isolated from sheep with pneumonia, as well as from healthy sheep, and at least one from goats. Sneezing, anorexia and pneumonia follow experimental inoculation of lambs, depending on the isolate used. The anterior lung lobes show collapse and dull red consolidation, the diaphragmatic lobes are turgid and rose-red coloured (7.1) and the pulmonary lymph nodes are enlarged. The lungs have a turgid or rubbery feel. The primary lesion is a proliferative bronchiolitis, with bronchopneumonia and oedema. Some isolates cause focal hepatic or renal lesions. Epidemiological surveys have shown that adenoviral infections are widespread among lambs, and persistent infections of up to 80 days can occur in older sheep.

Capripoxviral infection (sheep pox, goat pox)

The pulmonary form of these diseases is acquired through the respiratory system, or by viraemia from skin lesions (see Chapters 10 and 16). The multiple lesions of pulmonary consolidation (7.2) that result from such infection consist of proliferative alveolitis and bronchiolitis with focal caseous necrosis. Intracytoplasmic inclusion bodies can be seen in alveolar septal cells. Mortality is high in sheep and goats with pulmonary lesions.

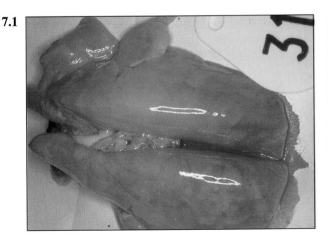

7.1

7.1 Adenoviral pneumonia. Experimental infection in a lamb. Note the oedema and congestion of the diaphragmatic lobes.

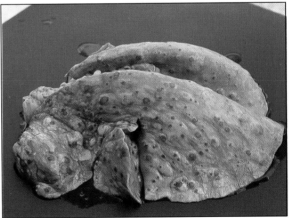

7.2

7.2 Sheep pox. Pulmonary lesions.

Herpesviral infections of sheep and goats

Pneumonia has been induced experimentally in germ-free lambs using intratracheal and intranasal inoculation of an ovine herpesvirus isolated from sheep with pulmonary adenomatosis in Scotland. Clinical symptoms are mild or absent, but at necropsy focal areas of dull-red consolidation are sometimes found in the lungs (**7.3**). These are due to interstitial pneumonia with marked histiocytic infiltration of alveolar septa (**7.4**), profuse exudation of lung macrophages, perivascular lymphoid cuffing, and occasional intranuclear inclusions in alveolar histiocytes.

Caprine herpesvirus (bovid herpesvirus type 6) has been isolated from nasal swabs and lungs of goats in New Zealand with fatal *Pasteurella haemolytica* pneumonia. Experimental inoculation of the virus alone produced only a mild catarrhal rhinitis and tracheitis. The role of these viruses in initiating pneumonia is uncertain.

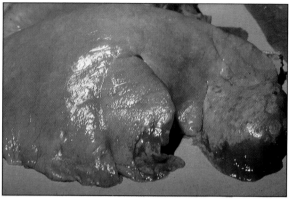

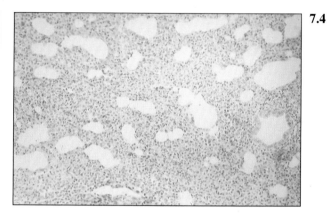

7.3 Ovine herpesviral pneumonia. Experimental infection showing focal consolidation.

7.4 Ovine herpesviral pneumonia. Marked histiocytic reaction in the lung. Haematoxylin and eosin.

Parainfluenza-3 (PI3) infection

Infection with parainfluenza virus type 3 (parainfluenza-3 or PI3 virus) causes mainly mild respiratory disease, but on occasion acute infections can cause severe morbidity in lambs under one year old. Such lambs are usually afebrile but cough frequently and may have a serous nasal discharge. Their lungs contain linear or sometimes confluent areas of consolidation and collapse, affecting mostly the apical, cardiac and the most anterior parts of the diaphragmatic lobes (**7.5**). Pleurisy is not a feature.

The disease starts as a mild, necrotising bronchiolitis, with collapse of adjacent lung and a mononuclear cell infitrate. After about one week, eosinophilic intracytoplasmic inclusions can be found in bronchiolar lining cells (**7.6**), and there is a transient proliferation of type 2 alveolar lining cells close to affected bronchioles. PI3 viral infections are worldwide in distribution, and represent a predisposing factor in some outbreaks of pneumonic pasteurellosis in sheep and goats.

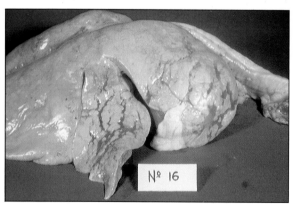

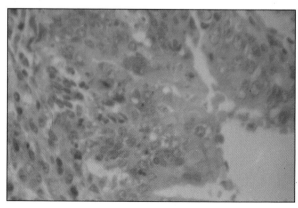

7.5 Parainfluenza-3 infection. Linear consolidation and collapse in lamb lungs.

7.6 Parainfluenza-3 infection. Acidophilic intracytoplasmic inclusions in the bronchiolar epithelium. Pollack's trichrome stain.

Chronic virus infections of the respiratory system

Caprine arthritis–encephalitis pneumonia (caprine progressive pneumonia)

The caprine arthritis–encephalitis virus (CAEV) and the maedi-visna virus (see below) are closely related lentiviruses in the family Retroviridae. Some adult goats infected with CAEV have a clinical syndrome of progressive dyspnoea (**7.7**), sometimes with weight loss indistinguishable from maedi-visna. Lesions of interstitial consolidation can be seen radiographically (**7.8**). Necropsy findings are also similar to maedi-visna. The caudal lung lobes and/or, cranioventral lobes are swollen, greyish-pink and firm, and the mediastinal lymph nodes are enlarged. A secondary anteroventral bacterial bronchopneumonia is often superimposed.

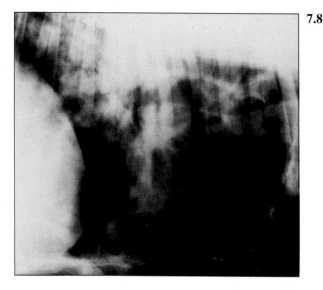

7.7 Caprine arthritis–encephalitis. Severe dyspnoea in the pulmonary form.

7.8 Caprine arthritis–encephalitis. Radiograph of lungs showing interstitial consolidation.

Maedi-visna (zwoegersiekte, ovine progressive pneumonia, la bouhite)

Maedi and *visna* are Icelandic words meaning respiratory distress and wasting, respectively. The disease is widespread throughout the world, though absent from Australia and New Zealand. The causal lentivirus is genomically distinct from CAEV, though sufficiently close for the two viruses to cross-react serologically.

Transmission is most commonly by the milk of infected ewes to their lambs. The onset of disease is insidious and, therefore, clinical signs of unthriftiness, inability to keep up with the flock and progressive dyspnoea are seen only in adult sheep. The lungs are the main target organ, but the central nervous system, joints, and udder may be affected (see Chapters 5, 6 and 8).

The lungs are grossly enlarged and very heavy compared with normal lungs. They are grey-brown in colour and often stippled with dark grey spots (**7.9, 7.10**). The cut surface has a solid, non-aerated appearance (**7.11**). All these effects are due to massive infiltration of alveolar walls by lymphoid cells, nodular lymphoid development and proliferation of smooth muscle. Infected sheep waste away and are culled or die of bacterial superinfections (**7.12**).

ELISA and/or AGID tests for antibodies to the virus are used in diagnosis of infection.

7.9

7.9 Maedi. The lungs fail to collapse and are stippled with tiny grey foci.

7.1

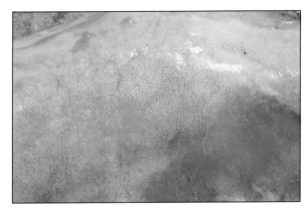

7.10 Maedi. Close-up of typical lung lesions.

7.11

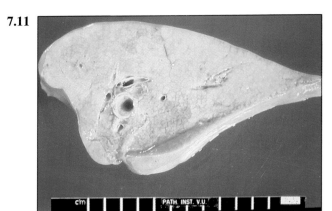

7.11 Maedi. Cross-section of a typical consolidated lung.

7.1

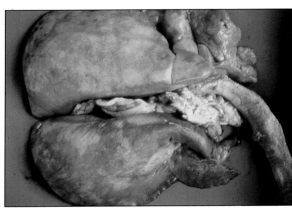

7.12 Maedi. Long-standing disease with secondary abscess formation.

7.13

7.13 Pulmonary adenomatosis. Copious nasal discharge in an affected sheep.

Pulmonary adenomatosis (SPA, jaagsiekte, ovine pulmonary carcinoma)

This disease is a transmissible lung tumour caused by a B-type or D-type retrovirus, possibly in association with a herpesvirus. Transmission is by inhalation of the fluid nasal discharge of an infected sheep that contains the virus. The incubation period is usually very long, often several years, though clinical disease can be seen in lambs under six months old. The disease is widespread in sheep, but less common in goats. No occurrences have been reported so far in Australia or New Zealand.

Affected sheep are often very lean in the autumn. As the disease progresses, they have difficulty in keeping up with the flock (*jaagsiekte* is Afrikaans for driving sickness) and exhibit abdominally assisted hyper-

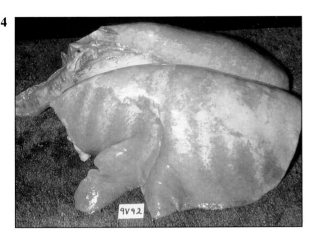

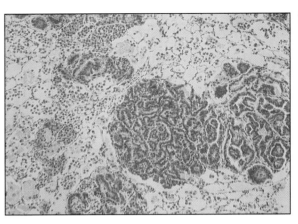

7.14 Pulmonary adenomatosis. Tumour consolidation showing the pattern of the ribs on the consolidated lung.

7.15 Pulmonary adenomatosis. Lung histology showing characteristic adenomatous foci. Haematoxylin and eosin.

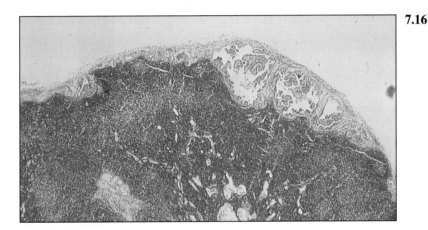

7.16 Pulmonary adenomatosis. Metastasis to a pulmonary lymph node. Haematoxylin and eosin.

pnoea when at rest. Auscultation reveals the presence of mucous rales. Fluid may pour from the nose if the hindquarters of the sheep are raised (**7.13**). The disease is progressive, and may terminate abruptly if the sheep becomes superinfected with *Pasteurella haemolytica* A biotype strains or other pathogenic bacteria.

The lungs are heavy and larger than normal due to the spread of tumour from discrete nodules to involve whole lobes or parts of lobes (**7.14**). Tumour tissue is solid, grey or light purple with a translucent sheen, often separated from normal adjacent lung by zones of emphysema. The cut surface often exudes a frothy fluid, which also fills the trachea. Complications include pleurisy with adhesions, abscess formation or secondary acute pneumonia, which may mask the tumour lesions.

The target cells in the lungs are the type 2 alveolar cells and the bronchiolar Clara cells. Proliferation of alveolar cells creates discrete adenomatous areas (**7.15**) which eventually become confluent. Thickening of the supporting fibrous stroma of the tumours often produces bizarre histological patterns. Large polypoid growths develop in, and cause partial occlusion of, bronchioles. Regional lymph nodes are hyperplastic and may contain metastases (**7.16**).

There is no serological test for pulmonary adenomatosis, as no antibodies to the virus are elaborated.

Pneumonias caused by mycoplasmas and bacteria

Contagious caprine pleuropneumonia

This economically important disease of goats occurs mainly in the Middle East, northern Africa and West Africa. Three organisms—*Mycoplasma mycoides* subsp. *capri*, *M. mycoides* subsp. *mycoides* (large colony type) and strain F38 (unclassified)—may be associated with explosive outbreaks of disease. The F38 strain is the cause of contagious caprine pleuropneumonia *per se*.

Caprine pleuropneumonia is highly contagious for goats but not for sheep. Infection is probably by droplet inhalation when animals are in close contact. Lesions are severe, with fibrinous or fibrinonecrotic pneumonia (**7.17, 7.18**), serofibrinous pleurisy and fibrinous pericarditis. Widening of interlobular septa and peribronchiolar interstitium by serofibrinous exudate can occur in some infections, though not when strain F38 is the infecting organism.

M. mycoides subsp. *mycoides* (large colony type) is widespread in North America and France, causing fibrinous polyarthritis with high mortality in kids, many of which also have fibrinous pleurisy and pericarditis, interstitial pneumonia and meningitis.

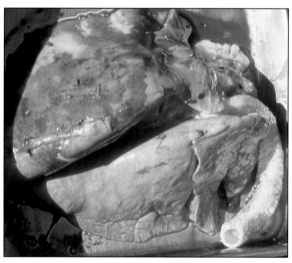

7.17 Contagious caprine pleuropneumonia. Acute lesions in the lungs.

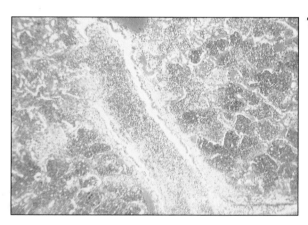

7.18 Contagious caprine pleuropneumonia. Lung histology showing acute exudative pleurisy and pneumonia caused by the F38 strain of *Mycoplasma mycoides* subsp. *mycoides*. Haematoxylin and eosin.

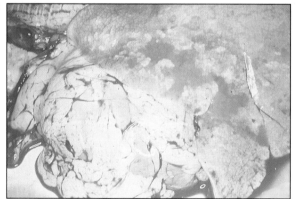

7.19 *Mycoplasma ovipneumoniae*. Experimental infection showing focal consolidation.

Mycoplasma ovipneumoniae infection

Although currently regarded as a facultative pathogen, *Mycoplasma ovipneumoniae* has been isolated, sometimes in large numbers, from cases of pneumonia in sheep and goats, especially where extensive peribronchiolar lymphoid cuffing has occurred. When given experimentally to specific pathogen-free lambs, *M. ovipneumoniae* induces a mild pneumonia with focal reddish consolidation of the anterior lobes of the lungs (**7.19**). In goats, experimental infection has induced fever and subacute fibrinous pleurisy.

Actinobacillosis

Infection of the subcutaneous tissues of the cheeks, nose, submaxillary and throat regions by *Actinobacillus lignieresi* (see Chapter 10) can spread to the regional lymph nodes, and occasionally to the lungs, in sheep. The lung lesions are large, yellow, encapsulated pyogranulomas (**7.20**). These contain multiple club colonies, or 'sulphur granules', in which masses of the coccobacilli are surrounded by radiating eosinophilic clubs composed of immune complexes.

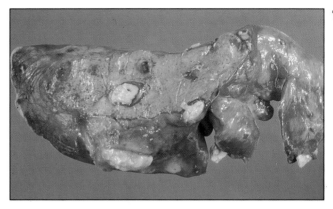

7.20

7.20 Actinobacillosis of the lungs in a sheep.

Caseous lymphadenitis

Chronic pulmonary abscesses occur in disseminated caseous lymphadenitis (see Chapter 14). These resemble the characteristic laminated lesions which normally develop in the lymph nodes (**7.21**).

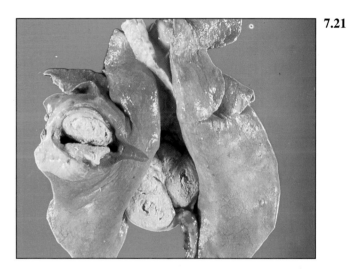

7.21

7.21 Caseous lymphadenitis abscesses in a sheep lung.

Melioidosis

Abscesses due to infection with *Pseudomonas pseudomallei* are occasionally found in small ruminants at slaughter in subtropical regions of Asia and in Australia (**7.22**). Such abscesses are not laminated, by contrast with those of caseous lymphadenitis in sheep, but culture is necessary to confirm the diagnosis.

7.22

7.22 Melioidosis abscesses in a goat lung.

7.23

7.23 Atypical pneumonia. Abattoir specimens with typical pneumonic lesions.

7.24

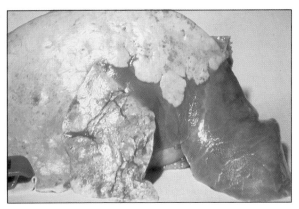

7.24 Atypical pneumonia. Lung showing sunken consolidation in the right apical lobe.

7.25

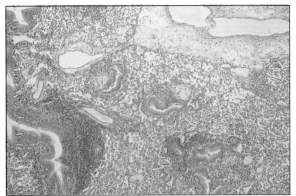

7.25 Atypical pneumonia. Lung histology showing collapse and hyaline nodule formation. Haematoxylin and eosin.

Non-progressive (atypical) pneumonia

This usually affects sheep under 12 months old, is often subclinical and rarely fatal. A similar but less common version of the disease can occur in goats. The disease is commonly recognised at the abattoir, especially during the winter (**7.23**). The cause is a mixed infection in which *Mycoplasma ovipneumoniae* and *Pasteurella haemolytica* biotype A serotypes predominate.

Experimental studies have shown that *M. ovipneumoniae* initiates an inflammatory reaction, which allows *P. haemolytica* to become established as a chronic infection.

Lesions are clearly demarcated areas of dark red or brown consolidation in the apical, cardiac and, less frequently, the anterior parts of the diaphragmatic lobes (**7.24**).

Alveolar collapse, infiltration by lymphoid and other mononuclear cells and extensive lymphoid cuffing of bronchioles and vasculature predominate. A diagnostic feature is the presence of nodular hyaline 'scars' close to bronchioles (**7.25**). Exudates of macrophages with variable numbers of neutrophil clusters can be seen in the compressed alveoli.

Outbreaks tend to occur in autumn, when batches of lambs from different sources are brought indoors for fattening. Travel, stress and overcrowding are also precipitating causes.

Pneumonic pasteurellosis (enzootic pneumonia)

The agent most commonly involved in pneumonic pasteurellosis is a gram-negative coccobacillus, *Pasteurella haemolytica* biotype A, of which 12 serotypes can be recognised by an indirect haemagglutination (IHA) test. Serotype A2 is the commonest found in both sheep and goats, and in sheep serotypes A1, A6, A7 and A9 also cause the disease.

P. haemolytica A biotype organisms are carried in the nasopharynx, and numerous factors, e.g. stress of dipping, castration, transport, etc.; change in feed, pasture or ambient temperature; and infection by viral agents, such as PI3, may be predisposing causes of pneumonia outbreaks.

Biotype A serotypes cause septicaemia in young lambs up to 12 weeks old, and pneumonia in older sheep. Young lambs are usually found dead with acute fibrinous pleurisy and pericarditis (**7.26**), but limited lung lesions. The myocardium, spleen, kidney and liver usually are petechiated, and the carcase lymph nodes are enlarged, soft and haemorrhagic.

In older sheep and goats, most cases are found dead. Live animals are dull, anorexic and have rectal temperatures above 40.5°C, rapid or distressed breathing, and nasal and ocular discharges. In hyperacute and acute cases, haemorrhages can be seen over the throat region and ribs when the skin is reflected. Subpleural and subepicardial petechiation is obvious, and varying amounts of clear, yellow pleural and pericardial exudate are found in the thoracic cavity.

In hyperacute pneumonia, lungs are heavy and cyanotic, with scattered bright purplish areas which exude frothy blood-tinged fluid when cut. The trachea and bronchi contain a frothy blood-tinged exudate, and their linings are engorged.

In animals which survive long enough to show clinical signs, lungs show extensive areas of dark red consolidation (hepatisation, **7.27**). The consolidated areas may be covered by a greenish pleural exudate, and may contain irregular dark brown or greenish-brown areas of necrosis (**7.28**).

In the chronic disease, the lungs may contain pink or light red consolidated areas, sometimes with patches of necrosis. Scattered nodular tumour-like masses are found in some animals. Surviving animals may have lung abscesses and/or, chronic pleurisy with fibrous adhesions.

The microscopic lesion in acute cases is characteristic. Necrotic areas have at their centres ghost-like remnants of the lung framework. Peripherally, wide zones of alveoli are filled with whorls of cells with flattened, basophilic nuclei—the so-called oat-cells (**7.29**)—which seem to stream from one alveolus to another. Masses of coccobacilli can be seen in the lesion and pleural exudate. Normally, diagnosis can be confirmed by cultural findings alone, but if colonies are sparse, histological and cultural findings are combined to confirm the diagnosis.

Outbreaks of pneumonic pasteurellosis are most common in the summer, with a second peak in late autumn. Secondary infections leading to chronic abscess formation are common (**7.30**).

Sporadic deaths attributable to *P. multocida* occasionally occur in sheep and goats. An acute anteroventral fibrinous pneumonia (**7.31**) is often found at necropsy.

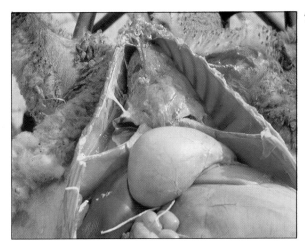

7.26

7.26 Pneumonic pasteurellosis. Acute fibrinous pleurisy and pneumonia.

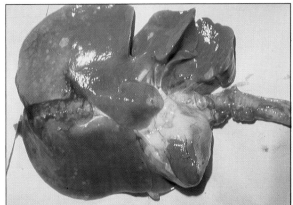

7.27

7.27 Pneumonic pasteurellosis. Hepatisation of the lungs in prolonged infection.

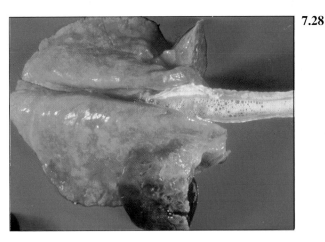

7.28

7.28 Pneumonic pasteurellosis. Acute necrotising pneumonia in the cardiac lobe.

7.29

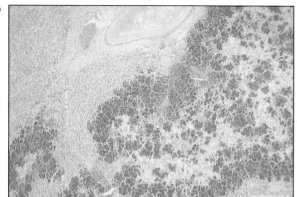

7.31

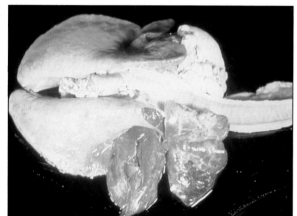

7.3

7.29 Pneumonic pasteurellosis. Focal necrosis with oat-cell response in alveoli. Haematoxylin and eosin.

7.30 Pneumonic pasteurellosis complicated by abscess formation.

7.31 Pneumonia in a goat, caused by *Pasteurella multocida*.

7.32

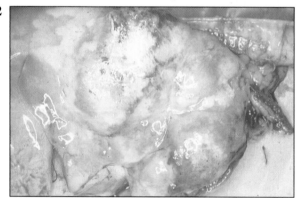

7.32 Chronic abscesses in sheep lungs.

Pulmonary abscesses

Lung abscesses occur in young lambs and kids mainly as a result of disseminated pyogenic infections, such as tick pyaemia or joint-ill (see Chapters 2 and 6).

In older animals, abscesses may develop as a consequence of chronic primary infection with pyogenic organisms. Animals with such abscesses may appear quite healthy. Additionally, abscesses may be found at necropsy as secondary findings in other chronic pulmonary infections; for example, pasteurellosis, maedi or pulmonary adenomatosis (see above).

Abscesses in the lungs are usually enclosed by thick fibrous capsules, while the surrounding lung is solid and rubbery **(7.32)**. Fibrous pleural adhesions often attach the affected portions of the lungs to the chest wall. Various pyogenic organisms may be cultured from lung abscesses, but a surprising number are bacteriologically sterile.

Systemic pasteurellosis

Lung lesions are often found at necropsy in sheep that have died of *Pasteurella haemolytica* biotype T septicaemia (see Chapter 2). The lungs are usually heavy and congested, with small dark-red or purple foci of consolidation. Microscopically, in hyperacutely infected animals, the lung capillaries contain numerous basophilic bacterial emboli (**7.33**). In acute cases, these emboli are niduses for local aggregates of flattened, basophilic leucocytes (oat cells).

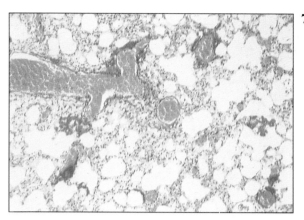

7.33

7.33 Systemic pasteurellosis. Bacterial emboli in lung capillaries. Haematoxylin and eosin.

Tuberculosis

Tuberculosis (see Chapter 2) is rare in sheep and goats. The route of infection is thought to be respiratory, because the thorax is the most common site for lesions. It is usually caused by *Mycobacterium bovis,* though occasionally by *M. avium* .

The lungs contain multiple caseous nodular foci which bulge from the pleural surface (**7.34, 7.35**). Similar lesions are found in the regional lymph nodes. Acid-fast organisms can be demonstrated in smears made from the cut surfaces of lesions.

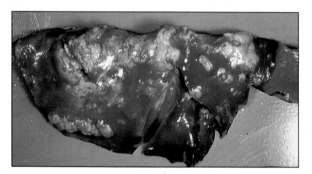

7.34 Tuberculosis. Pulmonary infection in a sheep.

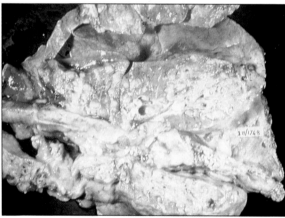

7.35

7.35 Tuberculosis. Lung cut in section to show tubercles.

Parasitic infections of the respiratory system

7.36

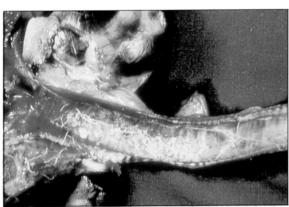

7.36 *Hartmannella* (*Acanthamoeba*) **sp. in sheep lung alveoli.**

Hartmannella infection

Pneumonia occasionally occurs in sheep in association with an amoeboid protozoon, *Hartmannella (Acanthamoeba)* sp. Infection probably occurs by inhalation, or regurgitation of water containing the protozoon. *Hartmannella* evidently can survive in the alveoli (**7.36**), and causes focal consolidation and alveolitis, with exudation of macrophages.

7.37

7.37 *Dictyocaulus filaria.* Adult worms in the trachea.

7.38

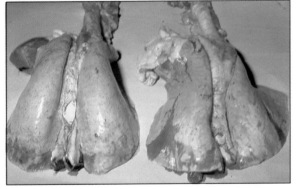

7.38 *Dictyocaulus filaria.* Typical sites of pneumonia in sheep lungs.

Lungworms

Parasitic bronchitis—'husk'—is common in young sheep in most environments, causing coughing, tachypnoea and unthriftiness. Goats are also susceptible to lungworm infections. Appetite may be reduced, with consequent weight-loss. The main cause is infection by the nematode *Dictyocaulus filaria*, the adults of which live in the trachea and bronchi.

The worm produces eggs, which hatch in the air passages. The larvae are coughed up, swallowed and pass in the faeces on to the pasture. When ingested by a new host, the larvae burrow through the wall of the small intestine and pass by way of the lymphatic system to the lungs. After moulting in the alveoli, the larvae finally pass to the air passages, the whole cycle taking about a month. Bronchitis usually occurs in the early autumn, and several cycles may take place. Larvae can overwinter, and infection can be carried over by chronically infected sheep or goats.

At necropsy, the adult worms are readily seen on opening the trachea and main bronchioles (**7.37**). Pneumonic areas may be scattered throughout the lungs, but the most common site is the posterior tips of the diaphragmatic lobes (**7.38**).

Severe hyperplasia and inflammatory changes take place in the air passages (**7.39**), with infiltration of the mucosa by neutrophils and eosinophils. Aspiration of eggs into the alveoli causes focal pneumonic consolidation and exacerbates the symptoms.

Both *Muellerius capillaris* and *Protostrongylus rufescens* have snails and other molluscs as intermediate hosts. The mode of infection and larval migration are similar to *Dictyocaulus*. Infections may be severe enough to cause outbreaks of coughing and unthriftiness in lambs in the autumn.

M. capillaris occupies the terminal bronchioles, alveolar ducts and alveoli, inducing the formation of characteristic greenish or greyish nodules on the dorsal aspects of the diaphragmatic lobes **(7.40)**. In some animals, localised zones of emphysema surround the parasitic nodules. Goats do not seem to develop immunity to *M. capillaris* and extensive lesions are often found in older animals. *P. rufescens* occupies the bronchi and small bronchioles, and infection is more widespread throughout the lungs **(7.41)**, with subpleural nodule formation **(7.42)**. Heavy infections cause severe emphysema **(7.43, 7.44)**. In both infections, stages of the parasite can be seen in histological sections **(7.45)**.

7.39

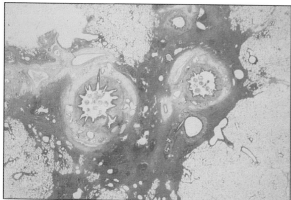

7.39 *Dictyocaulus filaria.* Parasitic bronchitis with adult worms in the bronchial lumens. Haematoxylin and eosin.

7.40

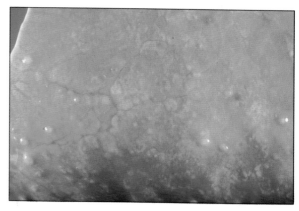

7.40 *Muellerius capillaris.* Nodular lesions in sheep lungs.

7.41

7.41 *Protostrongylus rufescens.* Focal lesions in sheep lungs.

7.42

7.42 *Protostrongylus rufescens.* Subpleural parasitic nodules in a sheep lung.

7.43

7.44

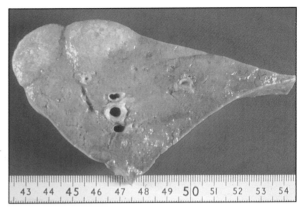

7.45

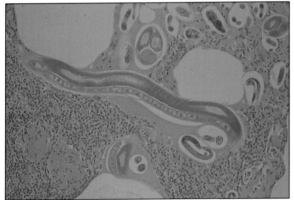

7.43 *Protostrongylus rufescens.* Verminous emphysema in a sheep.

7.44 *Protostrongylus rufescens.* Cross-section of an emphysematous lung.

7.45 *Protostrongylus rufescens.* Parasitic stages in the lungs. Haematoxylin and eosin.

Oestrus ovis infection

The nasal bot fly *Oestrus ovis* deposits its first-stage larvae on the nares. The larvae moult twice as they pass through the nasal passages, and eventually attach themselves by their mouth parts to the mucosa (**7.46, 7.47**). Their spinous cuticle causes irritation and a catarrhal rhinitis, with a nasal discharge. Those that penetrate into recesses in the sinuses and turbinates may be unable to leave after they reach maturity and die there. Affected sheep become debilitated but seldom die. Larvae sometimes penetrate to the cranial cavity and enter the brain (see Chapter 5).

7.46

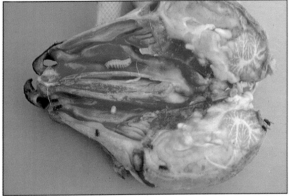

7.47

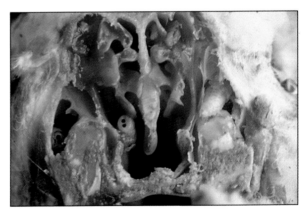

7.46 *Oestrus ovis.* Immature larval stage on the turbinate mucosa.

7.47 *Oestrus ovis.* Mature larvae infesting the nasal passages.

Miscellaneous diseases of the respiratory system

Aspergillosis

Among the less common complications of broncho-pneumonia is superinfection with the opportunistic fungus *Aspergillus fumigatus*. Colonisation of lung lesions by this fungus probably only occurs if animals are debilitated or immunodeficient. *Aspergillus* species are widespread in nature, but there may be situations where spores are in abnormally high concentration, e.g. on mouldy feed or bedding.

Affected lungs are consolidated and contain greyish- or greenish-yellow nodules (**7.48**).

A definitive diagnosis can only be established by histological examination using appropriate fungal staining methods. *Aspergillus* spp. hyphae are septate, branch dichotomously, and are circular in cross-section (**7.49**).

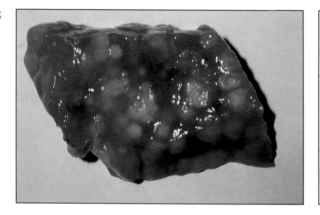

7.48 **Aspergillosis.** Nodular lesions in a sheep lung.

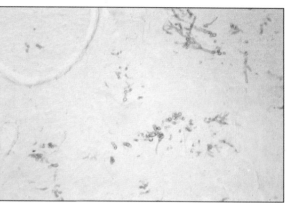

7.49 *Aspergillus* **spp. hyphae in a lung lesion.** PAS stain.

Atelectasis

This term means incomplete expansion of the lung. Fetal atelectasis occurs in stillbirths, where the neonate has never breathed. Such lungs do not float, and have a fleshy consistency (**7.50**). The lungs of stillborn animals are slightly blue due to dilatation of capillaries, whereas fetal lung is pink. If the neonate has breathed to any extent, pale pink aerated patches will be present.

Autolytic lung

The investigator must guard against hasty judgement where carcases are examined many hours after death. It is also important to consider the ambient temperature, as autolysis can be rapid in warm weather. It would be unwise to conclude that the lungs illustrated in **7.51** represent acute pneumonia, although superficially they resemble acute pneumonic pasteurellosis. Low counts of *Pasteurella* may be expected in normal lungs; thus cultural methods must be backed up by histology in doubtful cases.

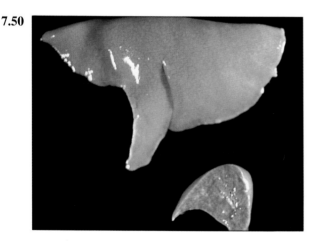

7.50 **Atelectasis.** Non-expanded lungs in a stillborn lamb.

7.51 **Autolytic lungs.**

Back bleeding

The lungs of slaughtered sheep and goats may contain large volumes of blood inhaled if the trachea and jugular veins are severed together when the carcase is bled out. Such lungs (**7.52**) may be wrongly assumed to contain focal areas of pneumonic consolidation.

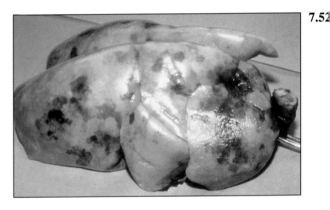

7.52 Lesions of back bleeding in sheep lungs.

Bronchiectasis

Bronchiectasis is dilatation of bronchi. It can be a sequel to chronic necrotising or suppurative bronchitis, where there is inflammatory weakening of the bronchiolar wall with accumulation of exudate. Usually the most dependent portions of the lungs are affected.

Externally, numerous nodular swellings protrude from the lung surface (**7.53**). When the lung is sliced, the dilated bronchi give the tissues a honeycomb or cystic appearance (**7.54**).

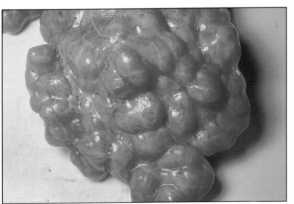

7.53 Bronchiectasis in a sheep.

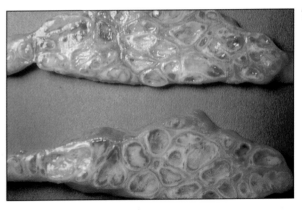

7.54 Cross-section through bronchiectatic lung, showing the dilated bronchioles.

Gangrenous pneumonia

The usual cause of this form of pneumonia is aspiration of foreign material containing putrefactive bacteria. Affected lungs are greenish or brownish (**7.55**) and distended with gas. The cut surface shows cavitation and necrosis, and the tissues have an offensive smell.

7.55 Gangrenous pneumonia in a sheep.

Ilex choke

Leaves of holly trees (*Ilex* spp.) are sometimes eaten by sheep, especially in the southwest of England, where the trees are numerous. The prickly leaves can sometimes clamp firmly, shiny surface uppermost, over the entrance to the larynx, the prickles making them impossible to dislodge (**7.56**). Affected animals die of suffocation.

7.56

7.56 Ilex choke. Holly leaf trapped over the laryngeal entrance.

Inhalation pneumonia

This condition is most common in recumbent animals which inhale their rumen contents and in orphan lambs or kids which are too weak to swallow properly when being force-fed. Improper drenching techniques, inhalation of plant awns (**7.57**), nutritional myopathy involving the muscles of deglutition, and regurgitation induced by toxic members of the rhododendron family (*Ericaceae*) are other causes.

Stomach contents or liquid feed often causes obstruction of the air passages with focal collapse, and bronchopneumonia is a common sequel. Microscopic evidence of vegetable matter and/or rumen protozoa in the bronchioles will confirm the diagnosis.

7.57

7.57 Plant-awn abscesses in a goat lung.

Lymphosarcoma

Nodular lesions of lymphosarcoma are occasionally found in the lungs at slaughter (**7.58**).

Melanosis

Irregular deposits of melanin are occasionally found in the lungs of lambs at slaughter (**7.59**).

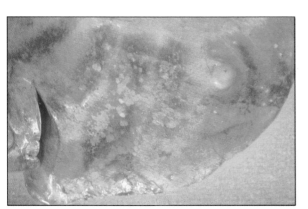

7.58

7.58 Lymphosarcoma nodules in a sheep lung.

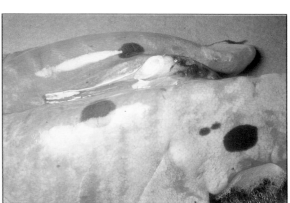

7.59

7.59 Melanotic pigmentation in the lungs.

Oedema (fog fever)

This condition is uncommon in sheep and goats, and must be distinguished from acute serofibrinous pneumonia. The cause of acute pulmonary oedema in small ruminants is assumed to be an acute anaphylactic reaction to inhaled antigens, antigens in the feed, or adverse reactions to vaccines. A relationship to fog fever in pastured cattle is not proven. Affected animals die rapidly and their lungs are waterlogged, with marked distension of interlobular septa, at necropsy (**7.60**).

7.60

7.60 Fog fever lesions in goat lungs.

Osseous metaplasia

This condition is occasionally found in old goats at slaughter, presumably following mineralisation of chronic pulmonary fibrosis (**7.61**).

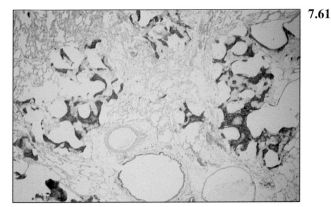

7.61

7.61 Osseous metaplasia in a goat lung. von Kossa stain.

Phenolic dip pneumonia

Outbreaks of acute, often fatal pneumonia have been described after dipping short-wooled breeds in phenolic dips designed to put a sheen on the wool for sale purposes. Some dips contain highly toxic fractions that are absorbed through the intact skin and excreted through the lungs. Affected sheep are dull and dyspnoeic, often with a distressed grunting form of respiration clearly audible from several metres. At necropsy, the carcase smells strongly of phenol and the lungs are solid and plum-coloured (**7.62**).

There is a strong macroscopic resemblance to acute pneumonic pasteurellosis, but lungs in phenolic poisoning are bacteriologically sterile and the histological appearance is pathognomonic. The type 2 alveolar cells are stimulated to proliferate and then shed in sheets into the alveoli (**7.63**), which become choked with cell detritus, proteinaceous fluid and inflammatory cells, while the alveolar capillaries are intensely congested. Replacement of type 2 cells by type 1 alveolar cells acts as a barrier to oxygen transport and causes death by suffocation.

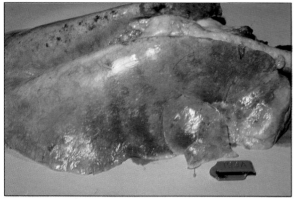

7.62 **Phenolic dip pneumonia.**

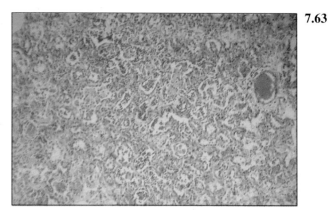

7.63 **Phenolic dip pneumonia.** Congestion and type 2 pneumocyte proliferation in the lung. Haematoxylin and eosin.

Pleural effusion

Large amounts of non-inflammatory fluid can accumulate in the thoracic cavity of sheep and goats; for example, in malnutrition, anaemia or congestive heart failure. Hydrothorax is also a feature of specific diseases, such as heartwater (**7.64,** see also Chapter 16), or black disease. Chronic hepatic disease with cirrhosis, and neoplasia of the thymus are other conditions in which hydrothorax can develop.

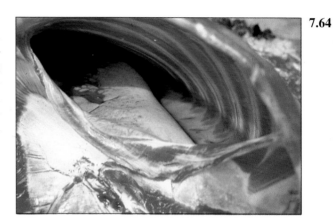

7.64 **Hydrothorax associated with heartwater.**

Post-dipping pneumonia

Pneumonia other than pasteurellosis can occur in sheep after dipping. This has been ascribed in part to inhalation of dipping fluid, with subsequent infection by bacteria in the fluid, or to opportunistic infection of stressed animals whose immunity to such infection is temporarily impaired.

The organisms involved are mainly streptococci, staphylococci, *Actinomyces pyogenes* or *Pseudomonas aeruginosa*. These can cause acute bronchopneumonia with anteroventral consolidation (**7.65**). In some animals, non-fatal infections progress to form chronic lung abscesses (see above).

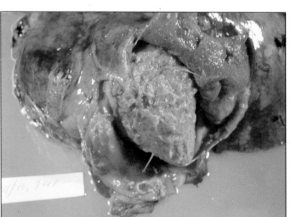

7.65 **Post-dipping pneumonia.** Consolidation of the lungs and pericarditis.

Shotgun wounds

Sheep and goats are shot occasionally by accident, usually in mistake for deer (**7.66, 7.67**).

7.66

7.6

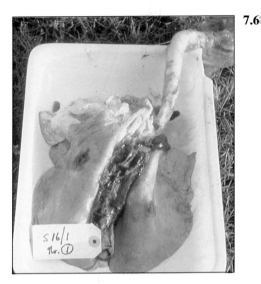

7.66 Gunshot wounds showing penetration of the thoracic wall.

7.67 Gunshot wounds in sheep lungs.

8 Diseases of the Reproductive System and Udder

This chapter illustrates many common diseases and disorders which can affect the male and female genital systems of sheep and goats, and which often interfere with normal mating or cause sterility. Diseases that affect the placenta and fetus are described separately in Chapter 9. Chapter 8 also deals with mastitis and other disorders of the udder, but lesions on the teats are covered in Chapter 10 and tumours of the mammary gland in Chapter 18.

MALE REPRODUCTIVE SYSTEM

Disorders of the Penis and Prepuce

Preputial adhesions

The normal prepubertal ram or buck cannot extrude its penis because of adhesions between the urethral process and glans penis, and the mucosa of the prepuce (**8.1**). Unless the animal is castrated at a young age, these adhesions gradually break down under the influence of testosterone, beginning first with the tip of the urethral process and then proceeding proximally.

8.1 Penis of a seven-week-old buck. Note the normal attachment of the urethral process and glans to the preputial mucosa.

Preputial orf

The parapoxvirus that causes orf (contagious pustular dermatitis, see **10.2–10.5**) can result in a venereal disease characterised by small pustules and shallow ulcers at the preputial opening of the ram or buck (**8.2**), and at the junction between skin and vaginal mucosa of the ewe or doe (see **8.43**). Some lesions are proliferative rather than ulcerated. Affected rams are reluctant to mate.

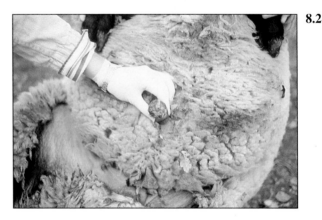

8.2 Preputial orf. Pustules and shallow ulcers at the preputial orifice of a sheep.

Posthitis (pizzle rot, enzootic posthitis, *C. renale* posthitis)

This disease principally affects castrated male sheep and Angora goats, particularly in Australia and Texas. The urease-producing diphtheroid *Corynebacterium renale* has been incriminated in many instances and can be cultured from the lesions. Excess dietary protein probably predisposes to infection. The skin of the preputial ring becomes necrotic and ulcerated, with extension to the preputial lining and penis (**8.3**). The prepuce becomes swollen and oedematous, and superficial ulceration develops on affected mucous membranes.

8.3 Enzootic posthitis. Swollen prepuce with areas of superficial necrosis.

Ulcerative balanitis, balanoposthitis

Contagious ulcerative lesions of unknown aetiology sometimes occur on the penis of the ram (**8.4**) and the vulva of the ewe (see **8.44**) during the breeding period. The ram retains libido, but copious haemorrhage from deep ulcers on the glans penis (**8.5**) may spread blood around the vulva of ewes in the flock.

The condition needs to be distinguished from injuries that occur when the ram is restrained on its rump and the glans penis is passively extruded during shearing (**8.6**).

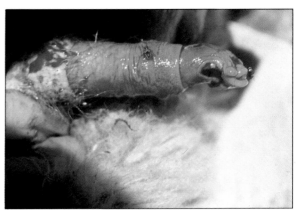

8.4 Balanoposthitis. Superficial necrosis involves the skin and mucosa of the prepuce and the glans penis.

8.5 Ulcerative balanitis of a ram's penis. Note the deep ulcer on the glans penis filled with a blood clot.

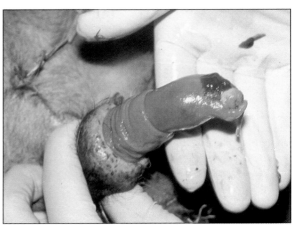

8.6 Penile injury. A healing wound, as from a shearing injury, is present on the glans penis.

Disorders of the epididymis, testis and scrotum

Epididymitis is a common cause of infertility in rams.

Brucella ovis epididymitis

The most widespread infectious agent in adult rams is *Brucella ovis* (the ram epididymitis organism or REO). This is transmitted between rams by homosexual behaviour, or when several rams mate with the same ewe, who acts as a fomite. Transmission can also occur across the mucous membranes of the mouth, nose or conjunctiva.

The organism localises in the seminal vesicles, bulbourethral glands and epididymis. If inflammation of the epididymis leads to ductal obstruction, the tail of the epididymis is initially swollen and tender, and then becomes palpably enlarged and firm (**8.7**). The testis may atrophy (**8.8**). Diagnosis is by semen culture (**8.9**) or serological tests (ELISA is preferred). Leukocytes and detached sperm heads in the semen are suggestive of *B. ovis* infection in rams which lack palpable lesions.

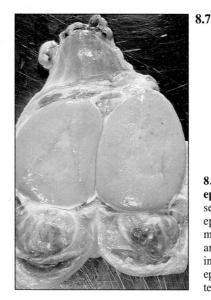

8.7

8.7 *Brucella ovis* **epididymitis.** Cross-section of testis and epididymis showing marked distension and fibrous thickening of the tail of the epididymis, while the testis is normal.

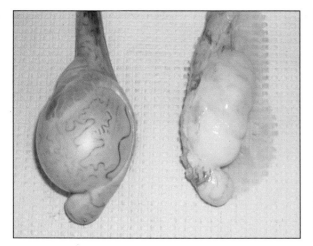

8.8 *Brucella ovis* **epididymitis.** The right testis is severely atrophied in this chronic infection.

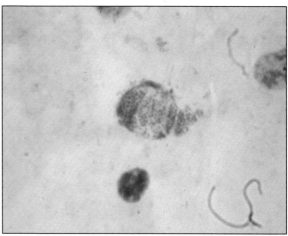

8.9

8.9 Seminal fluid with ruptured macrophage filled with red-staining *Brucella* organisms. Modified Ziehl–Neelsen stain.

Actinobacillus seminis and *Histophilus ovis* epididymitis

Gram-negative bacteria, variously identified as *Actinobacillus seminis* and *Haemophilus* or *Histophilus* spp. (lamb epididymitis organisms, LEOs), may be isolated in pubertal ram lambs and occasionally in adults. An acute epididymo-orchitis is accompanied by fever, depression and tenderness, and swelling of the scrotal contents (**8.10**). The tail of the epididymis is usually palpably enlarged (**8.11, 8.12**). Chronic lesions are firm and may drain pus through a fistula (**8.13**).

8.10

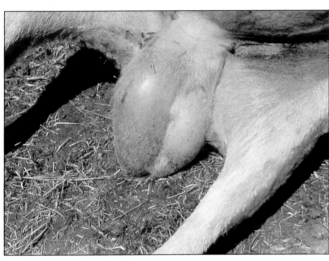

8.10 Lamb epididymitis in the scrotum of a ram lamb. Note the marked unilateral swelling and erythema. *Histophilus ovis* was isolated.

8.1

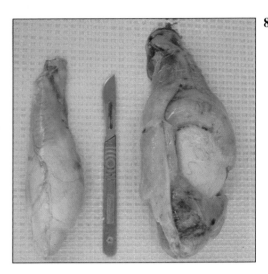

8.11 Lamb epididymitis in testes removed from a scrotum. Note the fibrosis and enlargement of the tail of the right epididymis due to *Histophilus ovis* infection. The left testis is atrophied.

8.12

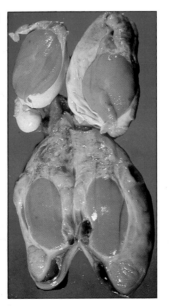

8.12 Lamb epididymitis in testes opened longitudinally. The affected epididymis (lower testis) is enlarged and the tunic is thickened. *Actinobacillus seminis* was isolated.

8.1

8.13 Lamb epididymitis in the scrotum of a five-month-old lamb. Swelling and crusts are associated with a draining fistula.

Orchitis

Orchitis is inflammation of the testis. This can result from trauma, or following infection—e.g. by *Actinomyces pyogenes* or *Corynebacterium pseudotuberculosis*. In bucks, orchitis can be caused by *Brucella melitensis*.

The lesion may be unilateral or bilateral, and is usually irreversible, leading to sterility (**8.14**). Even if unilateral, degenerative changes occur in the other testis. Initially, the testis is hot and painful and may be swollen, while chronic orchitis is characterised by reduced mobility and induration of the testis.

8.14

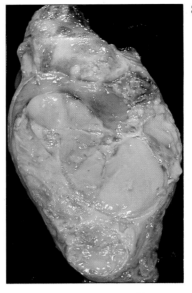

8.14 Orchitis and epididymitis. *Escherichia coli* was cultured from the abscesses destroying the parenchyma.

Spermatocele, sperm granuloma

Spermatocele results from obstruction of the epididymis, either as a congenital defect of epididymal tubules or as the result of chronic infection. As sperm are produced, the obstructed tubules become dilated and rupture, with release of spermatozoa into the interstitium, causing a granulomatous reaction. The affected epididymis feels nodular and is composed of encapsulated foci of milky or caseous material. In rams, the tail of the epididymis is most often affected, and the ram usually becomes infertile.

Hereditary sperm granulomas due to congenital obstruction of efferent ducts in the head of the epididymis occur in bucks, especially those that are homozygous for the polled trait (**8.15**). These bucks are usually initially fertile but become sterile if both epididymides become completely obstructed by granulomas. Testicular degeneration and mineralisation are frequent sequelae (**8.16**).

.15

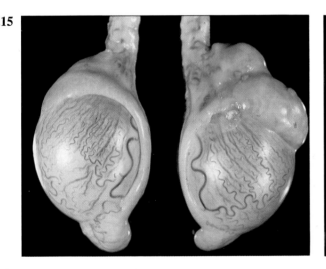

8.15 Sperm granulomas in a buck testes. The head of one epididymis is markedly and irregularly enlarged while the corresponding tail is shrunken.

8.16

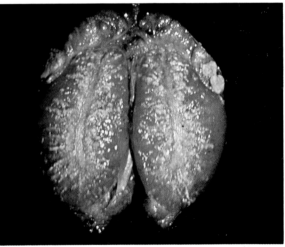

8.16 Sperm granulomas in a buck testis opened lengthwise. Note foci of mineralisation in the head of the epididymis and the parenchyma of the testis.

8.17

8.17 Testicular mineralisation in a buck testis opened lengthwise. Numerous white foci of mineralisation are present in the parenchyma of the testis, while the epididymis appears normal.

8.18 Besnoitiosis in a buck testis. Numerous parasitic cysts are present in the interstitium between the seminiferous tubules. Haematoxylin and eosin.

Mineralised testes

In aged bucks, it is common for sperm stasis and mineralisation to occur within the parenchyma of the testes bilaterally (**8.17**). The aetiology, in the absence of epididymal obstruction, is usually unknown. Trypanosomiasis is sometimes involved in Africa.

Besnoitiosis

Besnoitia besnoiti is a protozoal parasite occurring in Africa and the Middle East. The cat is the definitive host for the organism. Parasite cysts may be found in the walls and lumen of vessels in the pampiniform plexus as well as in the tunics, epididymis and testis (**8.18**). Thrombosis and testicular fibrosis lead to infertility.

8.18

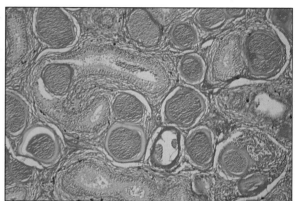

8.19

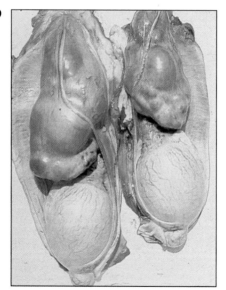

8.19 Varicocele. Extreme distension of veins in the pampiniform plexus bilaterally.

8.20 Varicocele. The neck of this ram's scrotum is enlarged and the scrotum almost touches the ground.

Varicocele

Varicocele is a unilateral or bilateral distension of the veins of the pampiniform plexus of mature rams (**8.19**). A firm and irregularly nodular swelling is palpable high in the neck of the scrotum, above the testis. If the varicocele is very large, the scrotum may touch the ground (**8.20**). Venous thrombosis is often present. Testicular degeneration and mineralisation may accompany the varicocele, and the affected ram is usually not judged to be a sound breeder.

8.20

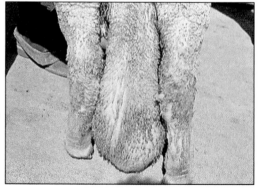

Inguinal hernia

Careful palpation will differentiate orchitis and epididymitis or varicocele from inguinal hernia (**8.21**), where fluctant intestines distend one side of the scrotum. The tendency for herniation is usually inherited.

8.21 Inguinal hernia. The right side of this Suffolk ram's scrotum is distended by herniated intestines.

Burdizzo infarction

Castration by the Burdizzo method involves crushing the blood vessels above the testis through the intact skin of the neck of the scrotum. Disruption of the blood supply leads to infarction and eventual resorption of the testicular parenchyma. The tunica albuginea and the epididymis remain viable if the artery accompanying the ductus deferens escapes the jaws of the castration instrument (**8.22**).

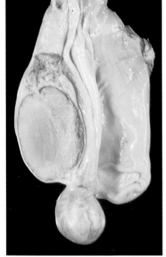

8.22 Burdizzo infarction of a sheep testis. The testis is shrunken and necrotic but the epididymis remains viable.

Testicular hypoplasia

This condition is an inherited trait, with either or both testes being smaller than normal (**8.23**). There is an accompanying failure of development of spermatogenic epithelium. Affected rams may be sterile or, if only one testicle is involved, of reduced fertility. Unilateral testicular hypoplasia has also been observed in bucks, but a genetic aetiology is unproven. Bilateral hypoplasia with no sperm production is one manifestation of the polled intersex condition in goats (see below). These 'bucks' are genetic females by karyotype, and are homozygous for absence of horns.

8.23 Testicular hypoplasia. One testis is smaller than the other.

Cryptorchidism

Failure of one or more testes to descend before birth (**8.24**) is an inherited problem that is most prevalent in Angora goats. The condition also occurs in rams; the prevalence in rams in the UK has been reported as 0.5%. The retained testis undergoes degenerative changes because of exposure to elevated temperatures within the body. The unilaterally affected male is fertile but should not be used for breeding.

8.24. Cryptorchidism. A single testis is present in the ram's scrotum.

FEMALE REPRODUCTIVE SYSTEM

Malformations

Freemartin

When placental fusion occurs and permits cells and hormones to pass from a male fetus to a female littermate early in gestation, the sexual differentiation of the female is perturbed. Variable externally visible abnormalities may be noted, such as the presence of an enlarged clitoris (**8.25**) or inguinal testes. The freemartin has XX-XY blood chimerism while its other body tissues have an XX karyotype. This permits differentiation from genetic intersexes associated with the polled condition (see **8.26, 8.27**).

Intersex

In goats of European breeds, the dominant gene for the polled trait is the same as or closely linked to a recessive gene for intersexuality. Homozygous polled female goats are sterile and have a variable phenotype indistinguishable from the abnormalities displayed by freemartins. Hypospadias (**8.26**) and clitoral enlargement (**8.27**) are common. Other animals appear as externally normal females or as males with scrotal, albeit hypoplastic, testes. Some intersexes have internal ovotestes. Intersexes usually act and smell like bucks during the breeding season.

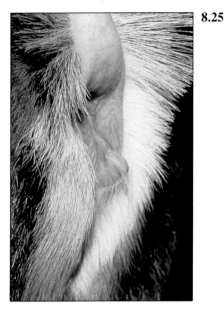

8.25

8.25 Freemartin. The vulva of this young Nubian goat with XX-XY blood chimerism is abnormally prominent.

8.26

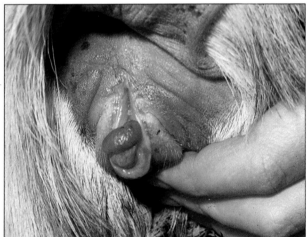

8.27

8.26 Intersex condition. Hypospadias and increased anogenital distance in a goat.

8.27 Intersex condition. An enlarged clitoris protrudes from the vulva.

The uterus and disorders related to pregnancy

Normal uterus and placenta

The normal midgestation placenta of sheep (**8.28**) and goats frequently has white foci on the chorion due to mineralisation and on the amnion due to keratinised epithelial plaques. These findings must not be construed as lesions. Ewes of black-faced breeds commonly have melanin deposited in caruncles (**8.29**), and this must be differentiated from necrotic lesions.

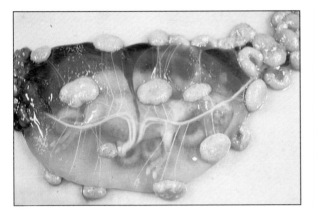

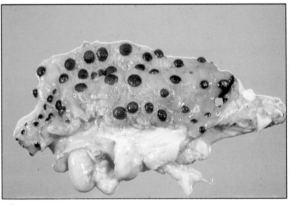

8.28 Normal fetus and membranes. White plaques on the chorion and amnion of this 122-day-old fetus are normal.

8.29 Normal melanosis in a uterus. Normal deposition of melanin in the caruncles of a primiparous Suffolk ewe.

Hydrometra, mucometra, pseudopregnancy

False pregnancy, often accompanied by accumulation of a mucoid but sterile fluid within the uterus, occurs rather frequently in unbred goats as well as in goats that are bred outside the normal season. Some does show marked abdominal distension, suggestive of pregnancy (**8.30**). If emptying of the uterus occurs spontaneously (termed a cloudburst) or is induced by administration of prostaglandin, the distension disappears (**8.31**) but the perineum is wet by the discharge. The diagnosis is made by demonstrating fluid in the uterus (**8.32**) in the absence of caruncles or fetus, as can be done by ultrasound examination (**8.33**).

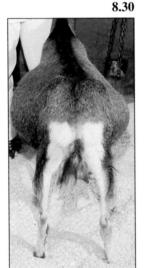

8.30 False pregnancy with marked abdominal distension.

8.31 Goat after correction of hydrometra. The abdominal distension is gone.

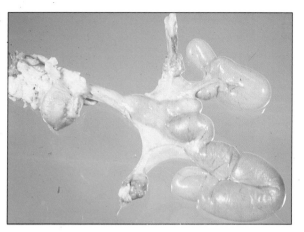

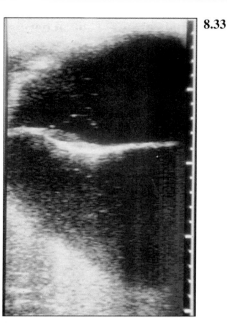

8.32 Hydrometra, goat uterus. The horns are thin-walled and distended with mucoid fluid.

8.33 Ultrasound scan of a goat uterus with hydrometra. The straight septum represents two adjacent sections of distended uterine horn. Slight flocculation is visible within the dark fluid.

Hydrops of the fetal membranes

Fetal anasarca can develop in either sheep or goats and may affect only one of twins (**8.34**). The extremely oedematous fetus may cause dystocia. When multiple lambs or kids are affected in a herd, a recessive gene is often implicated.

Hydrops amnion occurs when the amniotic cavity distends with excess fluid. The fetus is usually defective (**8.35**).

8.34 Anasarca. One kid is very oedematous while its twin is normal.

8.35 Hydrops amnion. The abdominal distension was the result of fetal abnormalities induced by Wesselsbron disease (see Chapter 16).

Abdominal rupture, ventral hernia

This condition usually arises from trauma and becomes progressively more obvious as late pregnancy ensues (**8.36**). Rupture may occur at the aponeurosis of the external and internal oblique abdominal muscles, or in the midline due to rupture of the pre-pubic tendon, where it inserts onto the pelvis.

An affected ewe experiences progressive difficulty in walking as pregnancy advances, and the ventral abdomen may almost trail on the ground, so that surgical assistance is required to deliver the offspring.

8.36

8.36 Abdominal rupture. The abnormally low abdominal floor is complicated by the distension of late pregnancy.

Vaginal prolapse

This occurs in the pregnant ewe or doe two to four weeks before parturition is due. It has been variously ascribed to hormonal or metabolic imbalances, over-fat body condition, bulky feed, lack of excercise and other less likely possibilities.

The prolapsed tissue becomes excoriated and this induces violent straining. If the bladder is involved in the prolapse, the straining is more severe. It is not unusual for prolapse of the rectum to occur simultaneously (**8.37**). If not replaced soon, the vagina may undergo necrosis, with resultant toxaemia. Loss of the cervical seal may lead to fetal infection and death, and sometimes to incomplete cervical dilatation.

If a tear occurs in the prolapsed vagina and straining continues, eventration of the intestines may occur, with death following rapidly (**8.38**).

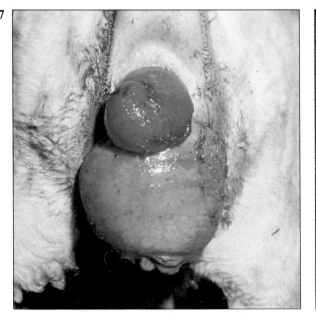

8.37

8.38

8.37 Vaginal prolapse in a pregnant ewe. The vagina and cervix are visible ventrally below the redder bulge of the prolapsed rectum.

8.38 Gut prolapse through the vagina. The ileum, caecum and part of the spiral colon are visible.

8.39

8.39 Dystocia in a goat. The front limbs are retained. The head is swollen because of impaired venous return.

8.40

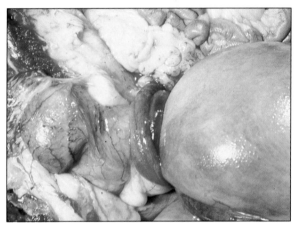

8.40 Uterine torsion in a ewe. The body of the uterus is red and oedematous where it is tightly twisted.

8.41

8.41 Uterine prolapse in a ewe. Caruncles are visible on the everted mucosal surface.

Dystocia

Dystocia is often due to fetal malpositioning, which is more likely to occur with multiples than with single fetuses. Retained head or limbs (**8.39**) and simultaneous presentation of two fetuses are relatively common events. Other causes of dystocia include fetal malformations, such as arthrogryposis; fetal oversize; uterine inertia related to pregnancy toxaemia, hypocalcaemia or fetal death; vaginal prolapse; and uterine torsion. Vaginal examination (except in dwarf goats) often permits identification of the cause of the dystocia. It may be difficult to distinguish between the first stages of labour in a normal animal and failure of cervical dilation (ringwomb) or uterine torsion in an animal with non-productive labour.

The cause of ringwomb is not understood, although hormonal imbalances are probably involved. The condition is seen more commonly in sheep than in goats. It is an occasional sequel to vaginal prolapse but also occurs independently of prolapse. The animal goes into labour but the cervix does not open, except perhaps enough to permit passage of one or two fingers or of part of the placenta.

Malpresentation of the fetus (breech position, retained head) leads to ineffectual straining and sometimes to spontaneous rupture of the uterus. Manual assistance during attempts to reposition a fetus or to dilate the cervix can also cause a full-thickness tear in the uterus, cervix or vagina. If infected fluid or placenta passes into the abdomen, septic peritonitis and death may result.

Uterine torsion

Torsion at the level of the vagina, cervix or body of the uterus (**8.40**) prevents presentation of the fetus into the birth canal. Straining is often minimal, but the animal may show initial acute abdominal pain followed by uneasiness or depression. Unless there is palpable involvement (spiralling) of the vaginal canal, this cause of failure to deliver cannot be differentiated from ringwomb without a laparotomy.

Uterine prolapse

Prolapse of the uterus (**8.41**) can occur within a few hours of parturition, particularly after a difficult parturition where vaginal damage or bruising has occurred. The fetal membranes are often still attached to the prolapsed uterine caruncles.

Metritis: rotten/emphysematous lambs

This condition may follow prolonged or unhygienic delivery, with introduction of harmful bacteria per vaginam. In cases where a ewe or doe in difficulty is unattended, the fetuses may die and become infected by anaerobic bacteria such as *Clostridium chauvoei*, with fetal emphysema and maternal toxaemia (**8.42**). Metritis may also follow retention of fetal membranes. Affected ewes and does have a foul-smelling bloody discharge and may be anorexic and febrile.

8.42

8.42 Emphysematous lambs. The uterus of this ewe contains decomposing lambs.

Infections of the external genitalia

Venereal orf

This condition is caused by a parapoxvirus similar to or identical with the orf virus. The most common site is the skin of the vulva (**8.43**), though ulcerative lesions can occur on the prepuce and penis (**8.2**). A similar or identical condition, ulcerative dermatosis (lip and leg ulceration), is associated usually with lesions on lips, nostrils, interdigital spaces and coronary bands of up to 60% of a flock. There is still some confusion as to whether these lesions have a common cause.

Venereal orf appears shortly after rams are turned out for mating. Small pustules form on the vulval labiae at the skin/mucosal junction, with subsequent ulceration. The ulcers become infected with *Fusobacterium necrophorum*. Rams become infected at the preputial orifice and are reluctant to mate.

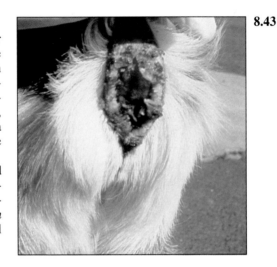

8.43

8.43 Venereal orf. Lesions on the vulva.

Ulcerative vulvitis

Vulvitis with swelling and ulceration of the vulva and posterior vagina has been associated with *Corynebacterium renale* and *Ureaplasma* infections. Short-docked ewes appear to be most susceptible.

A similar condition can occur in ewes two to three weeks after the start of mating (**8.44**), while associated rams have deep punched-out ulcers on the glans penis (**8.5**). The ulceration can give rise to severe haemorrhage from the glans, which is recognised when blood appears on the wool round the vulvae of the ewes. No infectious agent has been consistently isolated from such cases, nor have attempts at transmission been successful. The disease has to be differentiated from venereal orf. In goats, a similar condition has been ascribed to a herpesvirus.

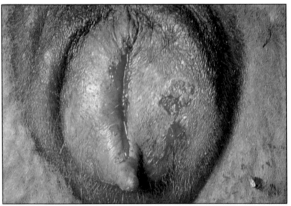

8.44

8.44 Ulcerative vulvitis. Note the marked oedema and reddened erosions. The aetiology is unknown.

UDDER

Malformations and anomalous lactation

Teat malformations are common in breeds that have not been selected for dairy purposes. They are discussed in Chapter 1 (see **1.39**).

8.45 Weeping teat. Milk can be expressed from pores in the skin at the base of this doe's teat.

Weeping teats

Acini of milk-secreting tissue are sometimes found in the wall of a teat. If this tissue communicates through skin pores to the outside, milk oozes onto the external surface of the teat, often near its base (**8.45**). If no duct is present, the milk accumulates in a nodule-like cyst within the wall of the teat. The nature of this lump can be determined by aspirating milk from it.

8.46 Witch's milk in a one-week-old kid. Note the drops of milk expressed from each teat orifice. Mammary development was induced by natural exposure to hormones *in utero*.

Lactation without pregnancy

Dairy goats are susceptible to hormonal induction of lactation and often exhibit mammary gland development and milk production without a preceding pregnancy. This can be seen as witch's milk in the neonate (**8.46**), precocious udder in the unbred doeling (**8.47**) and gynaecomastia in the buck (**8.48**). Factors favouring development of a precocious udder may include prolactin stimulation in the spring when day length is increasing and prolonged exposure to progesterone during a pseudopregnancy (see **8.30–8.33**). Males that develop functional mammary glands may have abnormal hormonal levels and become infertile but, otherwise, they are often normal bucks from lines with a genetic potential for high milk production.

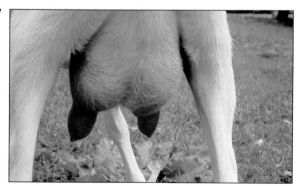

8.47 Precocious udder. Milk is present in this goat's udder even though the animal has never been pregnant.

8.48 Gynaecomastia. A buck with functional mammary glands.

Ectopic mammary gland

A functional mammary gland is occasionally found in the labia of a goat's vulva. A bilateral, firm, non-oedematous distension of this tissue first appears as the udder is enlarging shortly before parturition (**8.49**). The swelling gradually subsides during the three months after parturition as the secretory tissue undergoes pressure atrophy. Milk can be aspirated from the vulva to confirm the diagnosis.

8.49

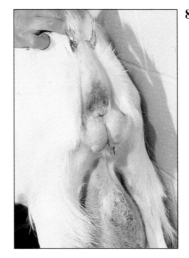

8.49 Ectopic mammary gland. This doe's vulva distends with milk each time it kids.

Mastitis

Bacterial and mycoplasmal mastitis

Mastitis is a common reason for culling ewes and does. Acute mastitis is commonly caused by *Staphylococcus aureus* and less often by *Pasteurella haemolytica*. Both of these can cause gangrenous mastitis (blue bag). Other causes of sporadic mastitis include *Escherichia coli*, *Streptococcus* spp., *Actinomyces pyogenes*, *Bacillus cereus*, *Clostridium perfringens* and *Pseudomonas aeruginosa*.

In the Middle East and North Africa, infection with *Mycoplasma agalactiae* is a common cause of mastitis (**8.50**), resulting in contagious agalactia in both sheep and goats. This disease appears shortly after parturition, when the udder may be hot and painful. Polyarthritis and keratoconjunctivitis may be present concurrently (see Chapters 6 and 12). Later, palpable nodules may develop,

followed by atrophy of the udder. Other mycoplasmas such as *M. mycoides* subsp. *mycoides* large colony type, *M. putrefaciens*, *M. arginini* and *M. capricolum* cause mastitis in goats. Many of these organisms also cause arthritis, pneumonia and keratoconjunctivitis. Heavy losses often occur in kids drinking unpasteurised milk.

Acute gangrenous mastitis commences with severe swelling of the udder with reddening and oedema of the skin (**8.51**), and the animal is usually lame. Peracute cases may occur in which death from toxaemia or septicaemia rapidly supervenes. However, the initial phase soon gives place to necrosis of mammary tissue, and the skin feels cold and clammy and has a purple hue (**8.52**), while the secretion is bloody and watery. Thrombosis is evident histologically (**8.53, 8.54**). The ewe or doe is ill,

8.50

8.50 Contagious agalactia. Acute mastitis in a goat, caused by *Mycoplasma agalactiae*.

8.51

8.51 Gangrenous mastitis. Oedema of the skin and discoloration of the udder parenchyma in a ewe with a clostridial infection.

8.52

8.52 Gangrenous mastitis. Note the purple discoloration of the udder and the limp teat of this Saanen doe.

febrile or not, and anorexic. Later, extensive tissue sloughing may take place, and healing may take weeks **(8.55, 8.56)**.

In acute non-gangrenous mastitis, the gland is swollen and firm, and the secretion is often watery or contains flecks or clots. There is, additionally, an increase in somatic cells in the milk.

Chronic mastitis **(8.57–8.59)** is characterised by nodular induration, often with abscess formation, from which *S. aureus*, *A. pyogenes* or *Corynebacterium pseudotuberculosis* may be isolated.

Warts and orf lesions at the teat end (see Chapter 10) predispose to bacterial mastitis.

8.53

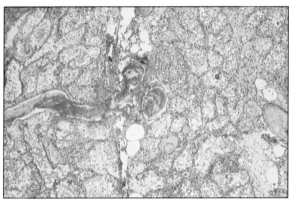

8.53 Gangrenous mastitis. Histological section showing fibrin thrombi in blood vessels and cellular exudate filling acini. Martius scarlet blue.

8.54

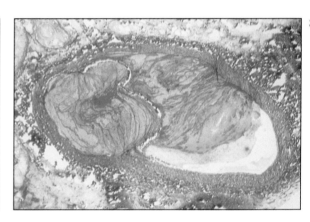

8.54 Gangrenous mastitis. Histological section showing a large, red-stained fibrin thrombus within an artery. Martius scarlet blue.

8.55

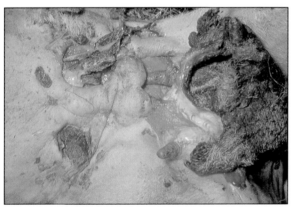

8.55 Gangrenous mastitis with sloughing. Note the red, depressed healing tissue in several areas and the large masses of dark, necrotic tissue that have not yet sloughed, two weeks after the onset of gangrene.

8.56

8.56 Gangrenous mastitis with sloughing now complete. The red fingers of tissue represent remnants of blood vessels encased in granulation tissue.

8.57

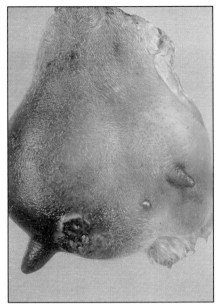

8.57 Chronic staphylococcal mastitis. The udder half on the left is firm and swollen and a draining fistula is evident.

8.58

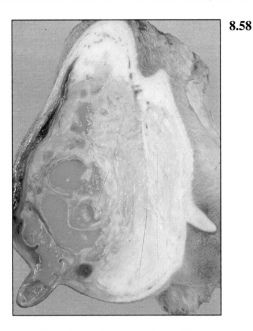

8.58 Chronic staphylococcal mastitis. The affected gland is distended with reddish liquid pus, while the other gland is atrophied.

8.59

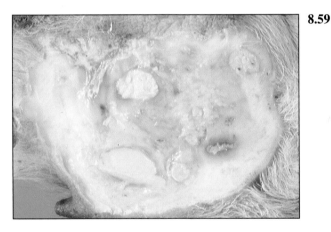

8.59 Udder abscess. This udder cross-section shows inspissated pus and accumulation of fibrous tissue around the abscesses.

Retrovirus interstitial mastitis

Caprine arthritis–encephalitis virus (CAEV) and maedi-visna (MV) or ovine progressive pneumonia virus (see Chapters 5, 6 and 7) are closely related retroviruses which are capable of causing lesions in the central nervous system, joints and lungs as well as in the udder. The viruses infect macrophages and are excreted in colostrum and milk. Signs of interstitial mastitis usually first occur at the time of parturition, when the udder is firm and distended but almost no milk can be expressed from the teats (**8.60**).

Diagnosis of infection can be made by serology or virus isolation, while definitive diagnosis of the interstitial mastitis requires histological examination and the ruling out of other infectious causes of mastitis. Normal goat or sheep milk, especially late in lactation, may contain large numbers of epithelial cells and macrophages (**8.61**). Demonstration of virus in milk macrophages from a CAEV-infected goat is not enough to confirm a viral aetiology for the mastitis.

The udder becomes diffusely indurated, causing progressive agalactia. There is extensive nodular lymphoid transformation, especially around ducts (**8.62–8.64**).

8.60

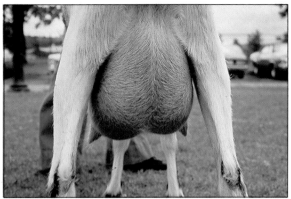

8.6

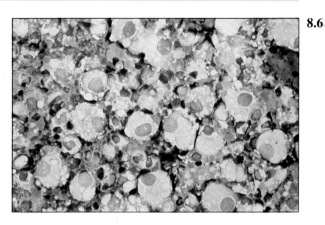

8.60 Caprine arthritis–encephalitis hard udder. This doe's udder was distended and very firm but no milk could be expressed.

8.61 Somatic cells in goat milk. Epithelial cells and macrophages are often present in large numbers.

8.62

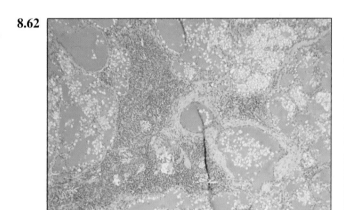

8.6.

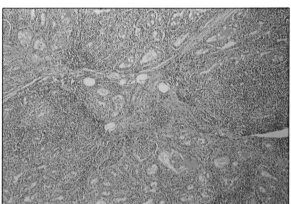

8.62 Caprine arthritis–encephalitis interstitial mastitis. Inflammatory cell focus within the interstitium. Acini contain clear milk fat globules and red milk protein. Haematoxylin and eosin.

8.63 Caprine arthritis–encephalitis interstitial mastitis. The parenchyma has been replaced by increased connective tissue and chronic inflammatory cells. Haematoxylin and eosin.

8.64

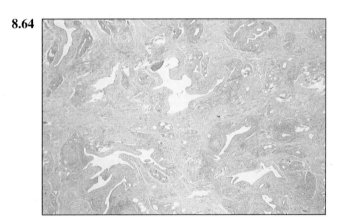

8.64 Maedi interstitial mastitis. Marked stromal thickening, loss of acini and periductal lymphoid foci. Haematoxylin and eosin.

9 Fetopathies and Diseases of the Placenta

Premature fetal death is an important cause of wastage and financial loss to the sheep and goat farmer. This chapter illustrates examples of the most common viral and bacterial infections which specifically cause placentitis and fetal death, albeit with little or no adverse effect on the health of the ewe or doe. Mention is also made of toxoplasmosis and of the fetopathic effects of a toxic plant. It must not be overlooked that, in addition to *Sarcocystis* spp., a number of acute systemic diseases—for example, Rift Valley fever, leptospirosis and salmonellosis—can cause abortion.

Fetopathic virus diseases

Akabane/Cache Valley virus infection

Bunyavirus infections in Australia, the Middle East and North America can cause abortions in sheep and goats. Transmission is by *Culicoides* midges and mosquitoes. Infection early in pregnancy can result in a range of deformities in the offspring, particularly microencephaly, hydrocephalus or hydranencephaly (**9.1–9.3**), accompanied by arthrogryposis (**9.4**) and severe muscle atrophy (**9.5**). The joint malformations may cause dystocia. Infection at a later stage may result in premature births and stillbirths.

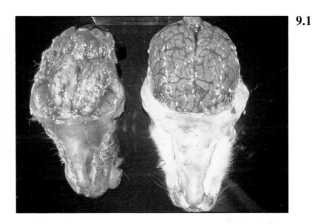

9.1

9.1 Akabane virus infection. A lamb brain showing microencephalopathy. A normal brain (right) for comparison.

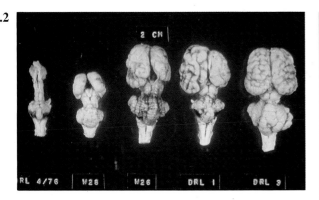

9.2

9.2 Akabane virus infection. Various brain defects in lambs.

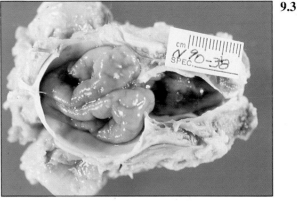

9.3

9.3 Cache Valley virus infection. Cerebral deformity in a newborn lamb.

9.4

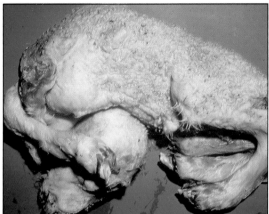

9.4 Akabane virus infection. A stillborn lamb with arthrogryposis.

9.5

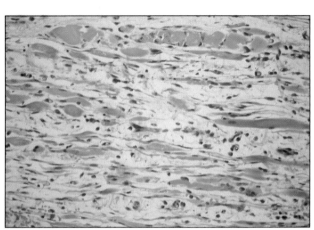

9.5 Akabane virus infection. Muscle atrophy.Haematoxylin and eosin.

9.6

9.6 Bluetongue. Cross-sections of fetal brains showing internal hydrocephalus or hydranencephaly.

Bluetongue

Bluetongue is an arthropod-borne orbivirus infection transmitted by biting midges (*Culicoides* spp.) which can affect both sheep and goats (see Chapter 2). Infection of pregnant animals at 50–55 days of gestation can cause abortions or fetal deformities, including hydrocephalus, hydranencephaly or porencephaly (**9.6**).

9.7

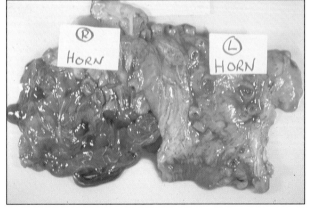

9.7 Border disease. Necrotising placentitis.

Border disease

Border disease is a congenital infection caused by a pestivirus serologically related to the bovine virus diarrhoea virus. Infection of the pregnant ewe can cause a spectrum of effects which are partly dependent on the stage of pregnancy in which infection was acquired. Infection causes a necrotising placentitis (**9.7**), which may rapidly lead to abortion or which may partly heal. Abortion can occur at any stage, but is most common around the 90th day of gestation, the fetus being mummified or anasarcous (**9.8**). Some stillborn infected lambs may have arthrogryposis (**9.9**).

Lambs born alive are generally infected as fetuses before the onset of immune competence. Such lambs, so-called hairy shakers (**9.10**), have muscular tremors of varying severity, and smooth-coated breeds have hairy and rough fleeces, with a halo of long hairs rising above the normal fleece, especially on the neck and back (**9.11**). Brown or black pigmentation may be present. The halo hairs are lost in a few weeks, but the fleece is always coarse and kempy. This is due to an increase in size of primary hair follicles with reduced numbers of secondary follicles.

The usual brain changes in hairy shaker lambs include myelin deficiency, an increase in type III glial cells and a reduction in oligodendroglial numbers. Such lambs are tolerant, persistent excretors of the virus. However, certain strains of the virus produce acute necrotising and inflammatory changes within the brain, including hydranencephaly and cerebellar hypoplasia or dysplasia (see Chapter 5).

In goats, border disease has caused abortion and shaker kids but has not been associated with changes in the birthcoat.

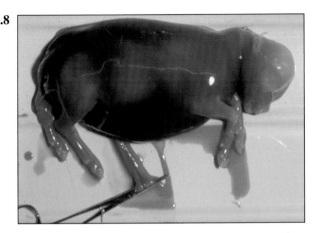

9.8 Border disease. Aborted fetus with anasarca.

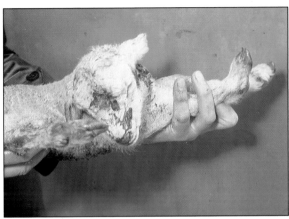

9.9 Border disease. Arthrogryposis in a newborn lamb.

9.10 Border disease. Hairy shaker lamb (foreground) with normal lamb.

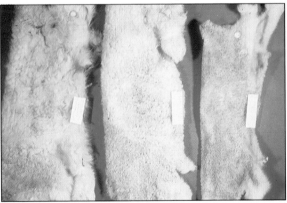

9.11 Border disease. Pelts of lambs of short-woolled breeds. The hairy shaker pelt is on the extreme left.

Wesselsbron disease

Abortions with fetal deformities (**9.12**) can occur in sheep infected in early pregnancy with a flavivirus related to the virus of Rift Valley fever virus (see Chapter 16). The virus is transmitted by *Aedes* mosquitoes.

9.1

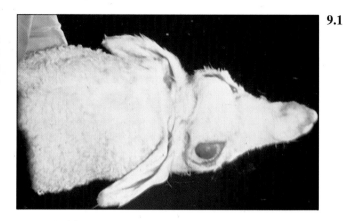

9.12 Wesselsbron disease. A fetus with craniofacial deformities.

Bacterial diseases resulting in placentitis and fetopathy

Brucellosis

Brucella melitensis is the principal cause of brucellosis of sheep and goats, being most prevalent in the Mediterranean region and the Near East. The disease is transmitted orally by ingestion of food contaminated by infected aborted fetuses or vaginal discharges. Infections can be fatal in goats but generally the disease is asymptomatic in both species. The organism may cause severe mastitis, especially in goats, and its excretion in the milk constitutes a serious human health hazard. Some strains of *Br. ovis* can cause abortion in sheep after sexual transmission and *Br. abortus* can cause sporadic abortions in sheep in contact with infected cattle.

After infection, a necrotising placentitis (**9.13**) slowly develops so that abortion occurs late in pregnancy (**9.14**). Retained placenta and metritis are associated with a prolonged, highly infective vaginal discharge. An allergic diagnostic test has been developed in some countries (**9.15**).

9.1.

9.13 *Brucella melitensis*. Necrotic cotyledons in aborted material.

9.14

9.15

9.14 *Brucella melitensis*. A stillborn fetus with necrotic placenta.

9.15 *Brucella melitensis*. A positive response to Brucellergen inoculated into the lower eyelid.

Campylobacter (vibrionic) abortion

This is caused by infection of pregnant animals with *Campylobacter fetus* subsp. *fetus* or *jejuni*. Infection in sheep is much more common than in goats. Abortion takes place late in pregnancy after a short incubation period, and lambs may be carried to term but are born dead and anasarcous, or alive but very weak. Infection causes a placentitis with arteriolitis and thrombosis in the hilar zones of the placentomes. Masses of cell-free organisms may be found at the tips of placental villi (**9.16**). The fetal liver is enlarged, haemorrhagic and, occasionally, studded with pale necrotic foci (**9.17**). Stillborn lambs may have fibrinous pleurisy and pericarditis (**9.18**). Diagnosis may be made by identifying the organisms in smears from cotyledons or fetal stomach contents, or by culture.

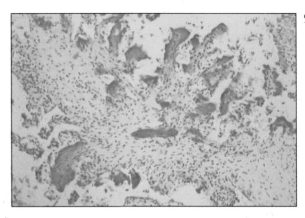

9.16

9.16 *Campylobacter* **abortion.** Cotyledonary villi packed with cell-free *Campylobacter fetus* subsp. *fetus* organisms. Haematoxylin and eosin.

9.17

9.17 *Campylobacter* **abortion.** Multifocal liver necrosis in a stillborn lamb.

9.18

9.18 *Campylobacter* **abortion.** Pleurisy and pericarditis in a stillborn lamb.

Chlamydial (enzootic) abortion

This disease, which is caused by the highly specialised bacterium *Chlamydia psittaci*, is one of the most common causes of sheep abortion in the United Kingdom, the Middle East and North America. Chlamydial abortion also occurs in goats. Contact with infected animals is a particular hazard for pregnant women.

The gut is the natural habitat of *C. psittaci*, but under certain circumstances in late pregnacy it can spread to the lymphatic system, causing systemic illness and infection of the placenta. The usual indication of an outbreak is when dead lambs or kids are discovered a few weeks before lambing is due. The dead animals are generally fresh, and may have blood-tinged fluid in their abdominal cavities. Fetuses may have necrotic or inflammatory foci in the liver and lymphatic system, skin and brain. Many stillborn or weakly premature animals may be delivered.

Typical lesions can be seen in the placenta after abortion. These are often oedematous or haemorrhagic, with necrosis of groups of cotyledons and leathery thickening of intercotyledonary areas (**9.19**), which may be covered by a cheesy pinkish exudate. The basic lesion is progressive necrosis of the trophoblastic cells and underlying chorionic villi of the placentome, associated with replication of the organism and the development of inflammatory changes.

Vast numbers of the organisms are present in infected placentas, and can be demonstrated in impression smears stained by a modified Ziehl–Neelsen stain (**9.20**), which stains the elementary bodies bright red. Immunological staining methods offer greater precision in diagnosis.

9.19

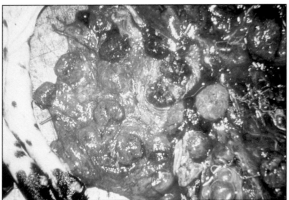

9.19 Chlamydial abortion. Typical intercotyledonary thickening in a sheep placenta, with yellowish surface deposits.

9.20

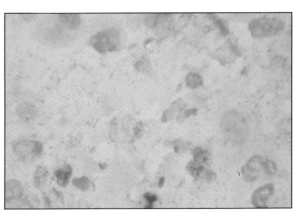

9.20 Chlamydial abortion. *Chlamydia psittaci* elementary bodies. Modified Ziehl–Neelsen stain.

Listerial abortion

Abortions resulting from infection of pregnant animals with *Listeria monocytogenes* generally occur in the last trimester of pregnancy. The route of infection is probably by ingestion. Infection of the placenta leads to acute placentitis (**9.21**) followed by septicaemia and death of the fetus. Fetal death may result in autolysis *in utero*; thus no recognisable lesions may be seen. However, in some fetuses multiple minute necrotic foci can be found in the liver (**9.22**) and, less commonly,

the spleen. If infection occurs very late in pregnancy, abortion may be complicated by dystocia, metritis and septicaemia in the dam.

The tips of the placental villi are usually necrotic, and profusely infected with *Listeria,* which can be seen on direct smear. Diagnosis is confirmed by culturing the organism from placenta, or from fetal liver lesions or stomach contents.

9.21

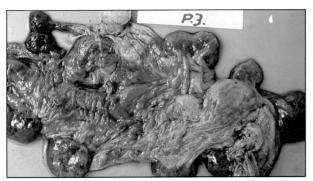

9.21 *Listeria* abortion. A thickened, leathery placenta.

9.22

9.22 *Listeria* abortion. Focal necrotic lesions in the liver of a stillborn lamb.

Q Fever

This disease is caused by the rickettsial agent *Coxiella burnetii*, which may be shed by infected parturient ewes or does in large numbers, though rarely causing abortion. However, a severe placentitis and abortion can occasionally occur (**9.23**). Vaginal discharges from infected animals therefore constitute a hazard for human contacts, as the organism can cause an acute, influenza-like disease, which may proceed to hepatitis or endocarditis. The organism can be demonstrated in uterine discharges and fetal stomach contents using a modified acid-fast stain (**9.24**). Human infections are diagnosed by measuring the serological response in paired serum samples.

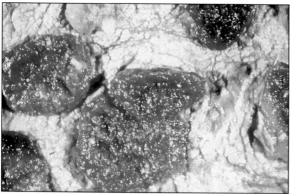

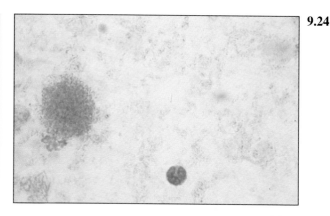

9.23 Q fever abortion. Placental lesions in an infected sheep.

9.24 Q fever: *Coxiella burnetii* **organisms.** Modified Ziehl–Neelsen stain.

Spirillum-like bacterial abortion

Flexispira rappini, an anaerobe isolated by Kirkbride and co-workers from aborted lambs in South Dakota, is a gram-negative curved tapered corrugated rod with up to 12 polar flagella (**9.25**). Abortions are late-term and sporadic. Focal hepatic necrosis in the fetus (**9.26**) resembles lesions usually ascribed to *Campylobacter* abortion (see above). The two diseases are differentiated by isolation of the causative agent using microaerophilic (*Campylobacter*) or strict anaerobic (*Flexispira*) conditions.

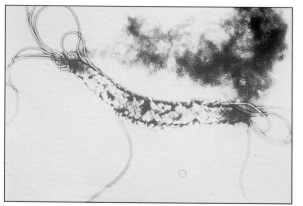

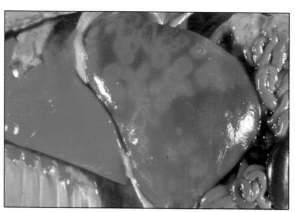

9.25 *Flexispira rappini.* An anaerobic abortifacient agent.

9.26 *Flexispira rappini* **abortion.** Necrotic liver lesions in a lamb fetus.

Miscellaneous causes of fetopathy or fetal death

Sheep–goat hybrid

Sheep and goats occasionally mate together. When a buck breeds with a ewe, fertilisation rarely occurs. When a ram breeds with a doe, a hybrid embryo develops. Immunological rejection of the hybrid placenta usually occurs by six weeks of gestation, resulting in resorption or abortion. Occasionally, the dead fetuses mummify and are not expelled until later (**9.27**).

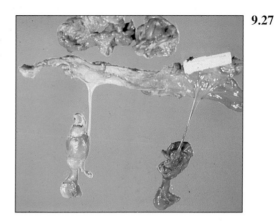

9.27

9.27 Sheep–goat hybrid. Fetuses from a doe mated to a ram.

Toxoplasmosis

Toxoplasmosis, a worldwide form of abortion that affects sheep and goats, is caused by the tissue-cyst forming protozoon *Toxoplasma gondii*, an intestinal coccidium of the domestic cat and other Felidae. An asexual phase of the parasite's life cycle takes place in most mammals, including wild rodents, sheep and goats. When these hosts are infected, tissue cysts containing bradyzoites form in the musculature and brain, without clinical disease.

In the pregnant ewe or doe, the organism infects the fetal membranes, with progressive damage to the cotyledons leading to infection of the fetus. If infection occurs in early pregnancy, fetal resorption may take place. Infection between 70 and 120 days of gestation causes abortions or stillbirths, or birth of undersized weakly lambs or kids.

Stillborn lambs may look quite fresh, but often are accompanied by a mummified fetus (**9.28**).

The placental cotyledons are dark red and speckled with white foci (**9.29**) that extend deeply into underlying tissue. These are focal areas of coagulative necrosis, often mineralised, which may contain toxoplasms. Necrotic lesions can also be found in the fetal brain (see Chapter 5), lung, liver, spleen, heart and kidney.

Serological tests have been developed to detect rising levels of anti-*Toxoplasma* antibodies in paired serum samples. Demonstration of *Toxoplasma* antigens using ELISA, or specific antibody to *Toxoplasma*, in fetal body fluids, using the latex agglutination or haemagglutination techniques, provides a diagnosis.

9.28

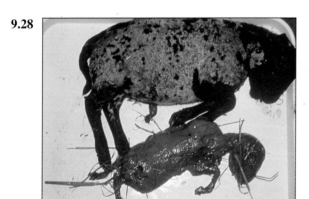

9.28 Toxoplasmosis. A mummified sheep fetus.

9.29

9.29 Toxoplasmosis. Typical focal necrosis of the cotyledons.

Veratrum californicum prolonged gestation

The consumption of *Veratrum californicum* (false hellebore) on day 14 of gestation leads to cyclopia and cebocephalic deformities (see Chapter 1). Some embryos die and are resorbed. Exposure to the plant on days 28–30 results in a shortening of the metacarpal and metatarsal bones; slightly later, the teratogenic effect is tracheal stenosis, which is fatal to the neonate. Another abnormality occasionally observed is prolonged gestation due to abnormal development or absence of the fetal pituitary. Affected lambs continue to grow *in utero* for an additional six to nine weeks. By then, their extreme size and possible deformity (**9.30**) cause dystocia and often death of the ewe.

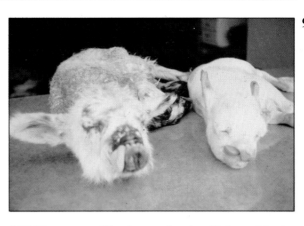

9.30

9.30 *Veratrum californicum* **poisoning.** Deformed lambs after prolonged gestation.

10 Skin and Ectoparasitic Diseases

The skin is the largest organ in the body, and has numerous protective and thermoregulatory functions. It is highly sensitive to systemic metabolic, immunological and hormonal irregularities, as well as being vulnerable to a wide range of external chemical, environmental and parasitic influences. Examples of skin infections, ectoparasitisms, metabolic diseases and reactions to trauma are illustrated in this chapter. Certain heritable skin conditions are covered in Chapter 1, and the effects of some viral infections that cause systemic illness with skin manifestations—for example, bluetongue and foot-and-mouth disease—are illustrated in Chapter 2.

Viral diseases

Capripox, sheep pox, goat pox

Both sheep pox and goat pox are caused by members of the genus Capripoxvirus of the family Poxviridae. The infection, which is contagious and often fatal, is described in Chapter 16.

Skin lesions begin as macules and evolve into raised circumscribed plaques or nodules, also termed papules. Vesiculation is minimal. Each nodule becomes hard and necrotic, with a depressed centre (pock). Eventually the scab detaches to leave a stellate scar. Lesions are most easily seen on the head (**10.1**) and external genitalia, but may be distributed over the entire dermis (see **16.1**). Histological examination of the lesions reveals vacuolated histiocytic cells with eosinophilic intracytoplasmic inclusions, cellules claveleuses or sheep pox cells, which are characteristic of the disease.

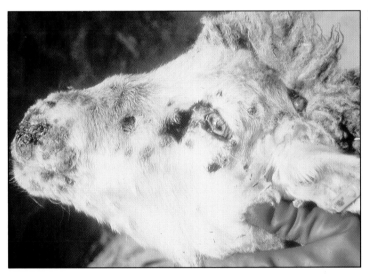

10.1

10.1 Sheep pox, adult sheep. Elevated crusts and scabs are prominent around the mouth and nares.

Orf, contagious pustular dermatitis, contagious ecthyma, sore mouth

Orf, a zoonotic viral disease of the skin caused by a parapoxvirus, occurs worldwide. Outbreaks are commonly seen in lambs and kids soon after birth or three to four months later. Older susceptible animals can become infected at any time.

Clinical signs are quite variable. The disease develops following abrasion or other trauma to the skin. Papules develop into pustules, which lead to scab formation, commonly at the commissures of the lips (**10.2**). Forcible removal of the scabs leads to bleeding. Lesions typically last two to four weeks. Proliferations of the dermis into warty or cauliflower excrescences sometimes interfere with feeding and often become infected with bacteria (**10.3**).

In some flocks, infection gives rise to a fulminating gingivitis and stomatitis (see Chapter 4); spread to the buccal cavity can cause high mortality.

Infection of the teats and udder (**10.4**) of ewes and does causes them to resent suckling. The overstocked udder becomes susceptible to infection, particularly with *Staphylococcus aureus* growing in the teat-end lesions, which can lead to gangrenous mastitis.

Limb lesions can be found on the thigh and in the axilla, but the most common site is at the coronet (**10.5**). In this site, haemorrhagic verrucose masses develop, giving rise to the term 'strawberry footrot'

10.2 Orf. Scabs at the commissures of a goat's lips.

10.3 Orf. Massive crusts involve the entire muzzle and eyelids of a sheep.

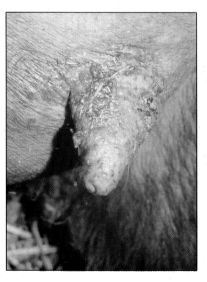

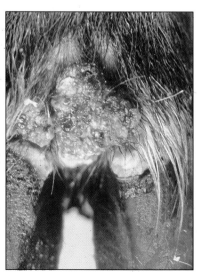

10.4 Orf. Erosions and scabs on the teats usually harbour secondary staphylococcal infections.

10.5 Orf. A proliferative lesion with scabs at the coronary band.

(10.6). Mixed infections with orf virus and *Dermatophilus congolensis* (see page 157) commonly occur in these sites.

Chronic orf lesions may be found on the poll (10.7) or external genitalia, and they can persist for months or years as unsightly scabby masses. Affected rams (and subclinical carriers) can introduce the virus to previously uninfected flocks.

Gloves should be worn when handling infected animals or live vaccine, as lesions on humans are painful and often persist for three or four weeks (10.8).

Venereal orf is discussed in Chapter 8, while foot-and-mouth disease, which can cause vesicular lesions at the coronary band (see 2.18) might be confused with orf or ulcerative dermatosis (see below).

10.6

10.6 Strawberry footrot. The combination of orf and *Dermatophilus congolensis* produces haemorrhagic, verrucose lesions.

10.7

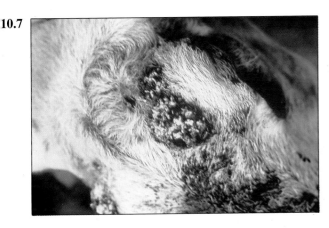

10.7 Poll orf. A chronic proliferative lesion on the top of a mature ram's head.

10.8

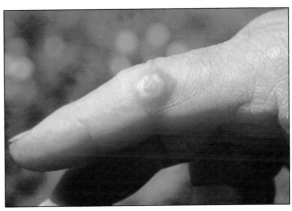

10.8 Human orf. Raised pustular lesion on a finger.

Ulcerative dermatosis, lip-and-leg disease

Though similar to orf virus infection, this disease is probably caused by a different virus. Circumscribed granulomatous ulcers develop on the hairy skin around the mouth (10.9) and on the legs and interdigital space, while the vulva or prepuce may also be involved. The classical lesions are common in sheep in South Africa and the western United States. The lesions are much less warty or verrucose than orf lesions. Infection is by direct inoculation into the epidermis.

10.9

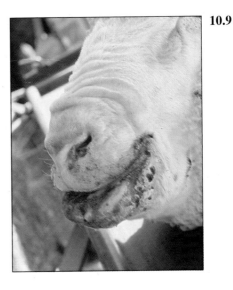

10.9 Ulcerative dermatosis. Ulcers on a ram's lips.

Bacterial diseases

Footrot

Footrot is a common and highly contagious disease of sheep, with a worldwide distribution; the disease also occurs in goats. The cause is a synergistic infection with the anaerobic bacteria *Dichelobacter nodosus* and *Fusobacterium necrophorum*. Factors such as moisture, breed and stocking density influence the onset of disease in a flock or herd.

The initial lesion is inflammation in the interdigital space similar to scald or scad (see page 155), and involves *F. necrophorum* alone. This organism invades

the skin at the junction with the horn, allowing virulent strains of *D. nodosus* to invade the deeper tissues (**10.10**). The hoof becomes progressively underrun, infection spreading often across the sole with separation of the horn from underlying tissues (**10.11–10.13**). Great pain and lameness result (**10.14**), and animals soon lose condition. The infection is accompanied by an offensive smell. Fly strike commonly occurs in warm weather (**10.15**).

10.10
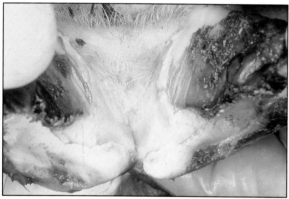

10.10 Footrot. Moist interdigital dermatitis and early underrunning of the sole.

10.1
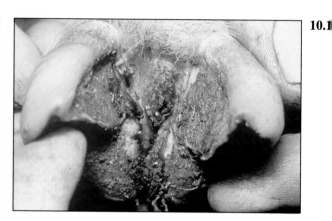

10.11 Footrot. The axial wall is blackened and necrotic.

10.12
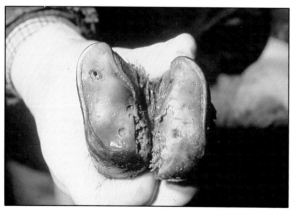

10.12 Footrot. The sole has been underrun and perforated in several places.

10.1
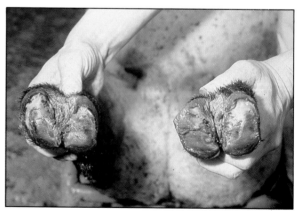

10.13 Footrot. The claws are deformed by a severe, chronic infection.

10.14 Footrot. This lame sheep is forced to graze on its knees as the infection spreads.

10.15 Footrot. Fly strike on the hoof of a sheep.

Scald, scad, interdigital dermatitis

This common condition can result from the action of almost any agent capable of causing inflammation of the interdigital skin. However, many outbreaks result from warm moist underfoot conditions that cause denaturation of the skin, thus encouraging infection with *F. necrophorum* or benign strains of *D. nodosus* (**10.16**). Scald is most common in lambs, although ewes are also often affected. Animals may be lame in several feet and resent handling for footparing. The sole is not underrun, however, and the offensive smell typical of footrot is absent.

Actinobacillosis

Superficial indolent granulomatous lesions occur in sheep when facial wounds become infected with *Actinobacillus lignieresi* (**10.17**). Secondary infection can result in abscess formation and suppurative adenitis in regional lymph nodes or lungs (see Chapter 7). The irritation associated with the lesions causes regional hair loss and superficial abrasions due to rubbing.

10.16 Foot scald. Interdigital dermatitis is not accompanied by underrunning of the sole.

10.17 Actinobacillosis. Pus exudes from granulomatous facial lesions.

10.18

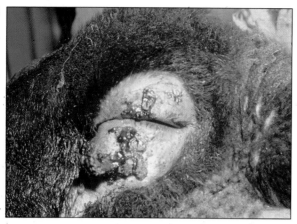

10.18 Periorbital eczema. Deep necrotic ulcers are present on the eyelids of a sheep.

Staphylococcal dermatitis

This is a pyoderma caused by a coagulase-positive, haemolytic *Staphylococcus* which produces alpha, beta and, possibly, delta toxins. Adult ewes are often affected just before lambing. The disease is highly contagious and is spread by fighting or jostling at troughs, especially if trough space is too restricted.

In its severest form (periorbital eczema, **10.18**), black scabs overlie deep necrotic ulcers which bleed easily. The skin over the nasal and maxillary bones, around the eyes, and at the base of the horns may be affected. Lesions occasionally develop on the feet (**10.19**) or on the teats.

Staphylococcal dermatitis in goats (**10.20**) is often generalised over much of the body, but secondary staphylococcal infection is a common feature of many skin diseases. Thus, diagnosis requires not only culture but also the elimination of other causes.

10.19

10.19 Staphylococcal dermatitis. A chronic ulcer on the fetlock.

10.20

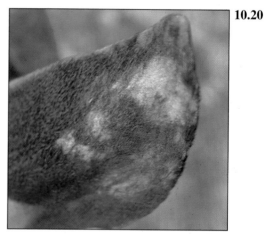

10.20 Staphylococcal dermatitis. Alopecia and scaling on a goat's ear.

10.21

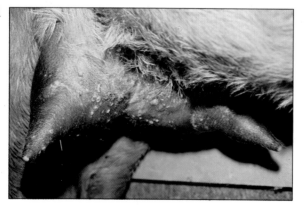

10.21 Staphylococcal folliculitis on the udder. In addition to the white pustules on the hairless areas, there is a moist dermatitis between the udder halves.

Staphylococcal folliculitis

Small, transient pustular lesions, mainly on the peri-abial and perineal skin of young animals, are caused by follicular infection with beta-haemolytic, coagulase-positive staphylococci. The crops of pustules seem to be more common in females, especially in the perineal region. They cause minimal distress and heal within a few weeks, although further crops may develop subsequently. Lesions on the udders of goats (**10.21**) are easily spread to others by milking procedures and predispose to staphylococcal mastitis.

Dermatophilosis, streptothricosis

This is a contagious zoonotic skin infection common in sheep and goats and caused by a gram-positive facultative anaerobe, *Dermatophilus congolensis*. The motile zoospores of this organism invade and proliferate in the epidermis under moist conditions.

The clinical effect is usually a low-grade scaly dermatitis (**10.22**), often diffuse along the back and extending downwards over the flanks. Detached scabs thicken and harden in the wool of sheep, binding fibres together; hence the term 'lumpy wool' (**10.23**). An equivalent form is occasionally seen on the back of goats (**10.24**). In moderate cases seen in temperate climates there is little or no pruritus, but in severe cases in warm climates the sheep rub themselves continuously, with wool loss and raw abrasions which attract blowflies.

A form mainly confined to young sheep and goats is localised to the ears or nose. In these sites, papule formation with exudation and multiple scabs causes a chronic dermatitis, and secondary infection can result in a painful eczema. It is probable that these chronic ear infections are the means by which the infection spreads in a flock. Lesions near the mouth or on the lower limbs between the coronet and carpus or tarsus (see under strawberry footrot, **10.6**) can be complicated by orf infection. In some cases, lesions on white areas appear to be complicated by photosensitisation (**10.25**).

Diagnosis can be confirmed by staining impression smears made from the undersurface of fresh scabs with Gram's or Giemsa's method, which demonstrates the characteristic branching filaments and coccoid forms of *Dermatophilus* (**10.26**). The organisms also fluoresce in ultraviolet light after staining with acridine orange.

10.23 Dermatophilosis. Wool fibres matted together by exudate; hence the term 'lumpy wool'.

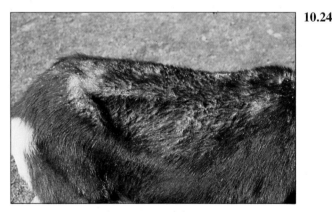

10.24 Dermatophilosis. Thick crusts cover this goat's dorsum after prolonged exposure to rain.

10.22 Dermatophilosis. Crusty, scaly lesions involve this goat's head, ears and neck.

10.25

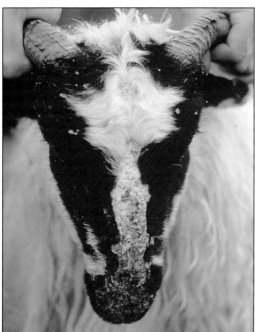

10

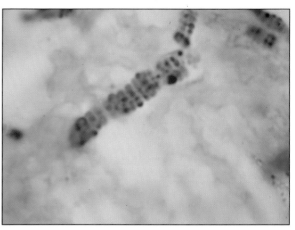

10.25 Dermatophilosis. Restriction of lesions to white areas suggests that infection induces photosensitivity.

10.26 Dermatophilosis. This smear from the underside of a crust shows the parallel arrangement of the organisms. Gram stain.

10.27

10.27 Fleece rot. Wool discoloration caused by yeasts and moulds.

Fleece rot

In Australia and New Zealand, *Pseudomonas aeruginosa*, a normal inhabitant of the fleece, is activated by moisture during rainy periods. Close-woolled sheep are at greatest risk. Proteolytic enzymes from the bacteria increase the superficial dermatitis and accumulation of the seropurulent exudate. The distinctive putrefactive odour attracts blowflies (*Lucilia cuprina*) to lay their eggs. The resulting fly strike can be fatal. Other opportunistic pigment-producing organisms produce a bright yellow (canary stain), greenish-blue (**10.27**) or brownish discoloration of the wool when a moist eczema develops.

Clostridial wound infection

Fighting among young rams occasionally induces contusions to the tissues around the horns. Infection of the wounds with *Clostridium novyi* results in oedema ('big head' or 'swelled head'), which must be differentiated from photosensitisation, and the subsequent toxaemia is rapidly fatal. Clostridia and other opportunistic organisms can cause severe cellulitis (**10.28**) after the vaccination of wet sheep or use of a dirty needle or contaminated product. Other vaccinal reactions are more localised, causing draining abscesses (**10.29**) or sterile granulomas.

Sternal abscess

Goats that spend much time in sternal recumbency, especially those with CAEV infection, may develop decubital lesions or draining fistulae (**10.30**) over the sternum. Sheep, especially rams, can acquire similar lesions. Abscesses are often surrounded by extensive fibrous tissue reaction (**10.31**). The presence of osteomyelitis of the sternebrae or spread of infection to lymph nodes within the thorax often contributes to chronicity and illthrift.

28

10.28 Cervical cellulitis. Painful neck as a result of vaccination.

10.30

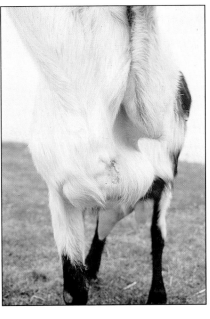

10.30 Sternal abscess. Firm swelling and draining track present over a goat's sternum for many months.

29

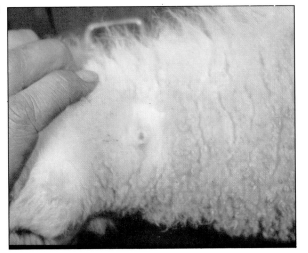

10.29 Vaccinal reaction. This draining lesion behind the ear originated from a caseous lymphadenitis vaccination.

10.31

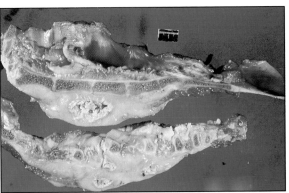

10.31 Sternal abscess. Cross-section of the sternum reveals pockets of exudate surrounded by dense connective tissue.

Fungal diseases

Ringworm

Ringworm, usually *Trichophyton verrucosum* or *T. mentagrophytes* infection, can occur in sheep and goats housed in pens that have previously held infected calves, although both species are fairly resistant to dermatophytoses. Infection is mainly on the face and ears and is usually acquired by rubbing against wooden partitions or mangers carrying the spores of the fungus. In some outbreaks, extensive lesions develop on the body, limbs and hindquarters (**10.32, 10.33**). Numerous other fungal species are occasionally isolated from ringworm in sheep and goats.

10.32

10

10.32 Ringworm. Wool loss and erythema caused by *Trichophyton mentagrophytes.*

10.33 Ringworm. Goat skin with alopecia and crusting.

Parasitic diseases

Chorioptic mange, foot mange

Chorioptes ovis and *C. caprae* are host-specific surface-dwelling mites which feed on epidermal debris. Lesions, which include nonfollicular papules, crusts, alopecia and erythema, are commonly confined to the lower limbs (**10.34**), inguinal area or scrotum (**10.35**). Severe cases in lambs are believed to lead to infertility. Pruritus, excoriations and secondary bacterial infections are typical.

There is an apparent allergic component to the disease in some goats. Other animals are asymptomatic carriers. Additionally, the mite can survive in the environment for up to 10 weeks. Diagnosis is by demonstration of the mite in skin scrapings or material collected with a flea comb. The mites have short, unsegmented pedicels and bell-shaped suckers on their pretarsi.

10.34

10.35

10.34 Chorioptic mange. Thick crusts on the lower limbs of a goat.

10.35 Scrotal mange. Crusting and alopecia on a ram's scrotum caused by chorioptic mites.

Demodectic mange

Demodex ovis and *D. caprae* are cigar-shaped mites which invade hair follicles and sebaceous glands. Palpable nodules form as mites, eggs and epithelial cells accumulate in the follicles. In goats, these nodules are most common on parts of the body exposed to friction (face, neck, shoulders, sides—**10.36, 10.37**). The nodules (**10.38**) are often pea-sized by the time the goat is 18 months old but usually remain painless and covered with normal hair. The toothpaste-like contents can be expressed for microscopic examination (**10.39**). Hide damage is extensive (**10.40**). Subclinical carriers harbour the mites in the skin of the eyelids, prepuce or vulva. Immune deficiency appears to contribute to the development of lesions in goats.

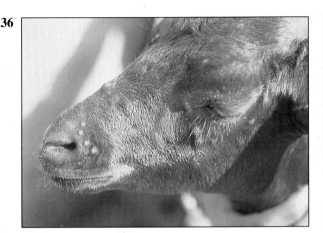

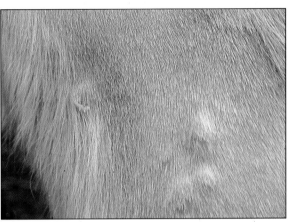

10.36 Demodectic mange. Nodules surrounding a goat's nares and eyes.

10.37 Demodectic mange. Nodules have been exposed by clipping the hair, and exudate has been expressed from one nodule.

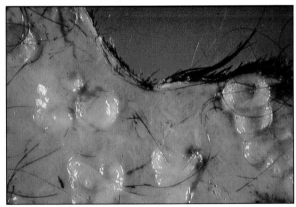

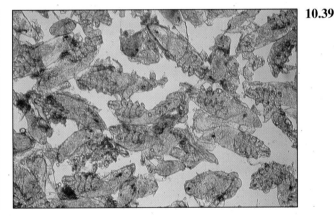

10.38 Demodectic mange. Nodules seen from the underside of the skin.

10.39 *Demodex* mites. Smear of exudate from a nodule.

10.40 Demodecosis. Damaged hides.

Psorergatic mange, psorergatic itch

Psorergates ovis (**10.41**) causes extremely pruritic but only slowly progressive alopecic, scurfy lesions (**10.42**) on the side, flank and thigh of sheep (especially Merinos) in Australia, South Africa and elsewhere. The mites are very small and difficult to detect.

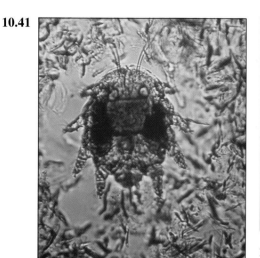

10.41

10.42

10.42 Psorergatic itch. Damaged fleece and skin.

10.41 *Psorergates ovis*. Female mite in skin scraping.

Psoroptic mange, sheep scab

This highly contagious disease, notifiable in the United Kingdom, results from infection with the mite *Psoroptes ovis*. This develops on the skin surface, the life cycle being complete in less than three weeks after initial infection. Multiplication takes place rapidly and an infected sheep is soon covered with lesions.

The mites pierce the epidermis to feed on lymph. Initially yellow pustules form, and a crusty exudate causes moistness and matting of the fleece. As the dried crusts part from the skin, the wool comes away with them (**10.43**), leaving bare patches (**10.44**). Goats are also susceptible (**10.45**). The irritation is intense, and infected sheep make 'mouthing' movements when touched, or roll on the ground in paroxysms. Animals will not feed, become emaciated and may die if not treated.

Diagnosis depends on accurate identification of the mite in skin scrapings. The mite has a long, three-segmented pedicel with a funnel-shaped sucker (**10.46**).

10.43

10.

10.43 Psoroptic mange. Tags of wool hang from this pruritic sheep.

10.44 Psoroptic mange. Yellowish crusts lift away from the skin.

10.45 Psoroptic mange. Alopecia and scales can be seen on this goat's head.

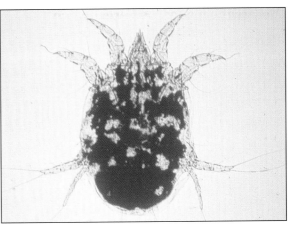

10.46 *Psoroptes ovis*. The pedicels are long and jointed, with characteristic funnel-shaped suckers.

Ear mites

Psoroptes cuniculi mites very commonly inhabit the ears of goats, where they induce head shaking and kicking at the ears. The mites are usually surrounded by thick wax, deep in the external ear canal **(10.47)**, and are difficult to visualise unless the goat is sedated. Occasionally, mites extend outwards onto the head. *Raillietia* spp. have also been found in goat ears. Ear mites may have a greater significance as carriers and transmitters of various mycoplasmas from animal to animal.

The deformed external ear that typifies the LaMancha goat breed may develop an otitis externa, with or without mites **(10.48)**. In sheep, head-shaking induced by the ear irritation can lead to haematoma formation.

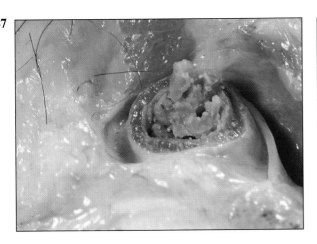

10.47 Ear mite infestation. Wax blocking goat's ear canal.

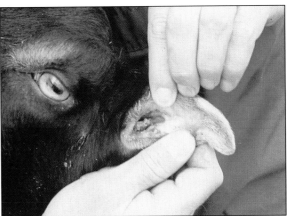

10.48 Otitis externa. A plug of exudate is present in this LaMancha goat's ear.

10.49

10.49 Sarcoptic mange. A goat with generalised lesions.

Sarcoptic mange

The strains of *Sarcoptes scabiei* are usually host-specific but can be transmitted between sheep and goats. Lesions, which are very pruritic, often begin on the head but may soon spread to the rest of the body, especially in goats (**10.49, 10.50**). There is hyperkeratosis and alopecia, and in advanced cases the skin is wrinkled and deeply fissured. Lesions on the lips and muzzle interfere with feeding. Animals lose weight and may die. The value of the hide is greatly decreased. Mites demonstrated in skin scrapings have long, unjointed pretarsi. Scrapings are often negative, but mites (**10.51**) or their tunnels may be found in biopsy specimens.

10.50

10.50 Sarcoptic mange. This goat has chewed on leg lesions.

10.

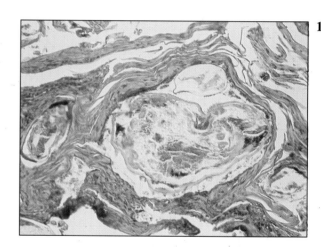

10.51 Sarcoptic mange. Cross-section of a mite in skin. Haematoxylin and eosin.

Louse infestation

Biting lice (*Damalinia* spp. (**10.52**), *Holokartikos crassipes*) and sucking lice (*Linognathus* spp., **10.53**) spend their entire life cycle on the host. Eggs are attached to wool fibres or hair shafts. All species cause irritation, but sucking lice also contribute to anaemia. Infestations are most severe in colder months.

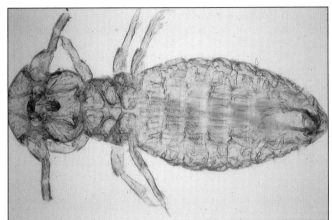

10.52 *Damalinia ovis*. This biting louse has a broad head and a tan body.

10.53 *Linognathus ovillus*. This sucking louse has a narrow head and a blood-filled body.

Melophagus infestation, keds

Melophagus ovinus is a wingless fly 4–6 mm long (**10.54**). The ked spends its entire life cycle on sheep or goats. It is brownish-grey and has a broad head with piercing mouthparts. The presence of only three pairs of legs distinguishes the ked from an adult tick (see **10.56**). It feeds by piercing the skin and ingesting blood. This annoys the animal, causing it to bite, kick and rub itself. Fleece and skins are thereby damaged and growth rate may be retarded. Severely infested animals may become anaemic.

Fleas

Sheep and goats do not harbour their own particular flea species, but housed young goats can become heavily infested with *Ctenocephalides felis* if these are numerous in their environment (**10.55**). The fleas cause considerable irritation, and infested animals cannot rest.

10.54 *Melophagus ovinus*. The sheep ked, a wingless fly.

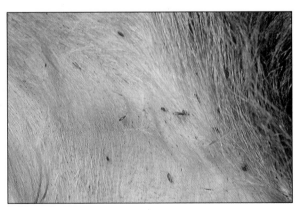

10.55 Ctenocephalid fleas from cats on the skin of a young goat.

10.56

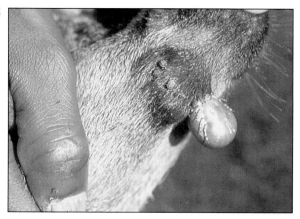

10.56 *Amblyomma hebraeum*. This engorged tick is attached below the chin.

10.57

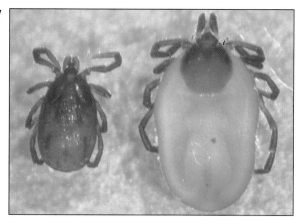

10.57 *Ixodes ricinus* ticks. The female is larger than the male.

10.58

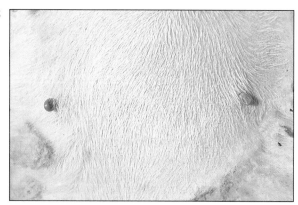

10.58 Ticks *in situ*. Two *Ixodes* ticks attached to a sheep's ventrum.

Ticks

Amblyomma hebraeum (**10.56**) is a three-host tick which transmits *Cowdria ruminantium*, the rickettsial agent which causes heartwater (see Chapter 16). Other *Amblyomma* species transmit heartwater in Africa and the Caribbean region.

Ixodes ricinus (**10.57, 10.58**), a three-host tick in Europe, is an important vector of louping ill (see Chapter 5), tick-borne fever and babesiosis (see Chapter 14). Ticks also cause irritation (tick worry), blood loss, damage to hides and tick paralysis.

Headfly myiasis

Harassment by the sheep headfly, *Hydrotoea irritans* (**10.59**), which is particularly active on still, hot summer days in Northern Europe, can result in self-mutilation of sheep, as they rub their heads on the ground or with their hind feet. The resultant abrasions, with seepage of serum or blood (**10.60**), render the sheep even more attractive to the flies. Horned breeds of sheep, and those whose faces are covered with hair rather than wool, are the most susceptible. The headfly lays its eggs on the ground, but bacterial infections or blowfly strike may complicate the skin lesions.

10.5

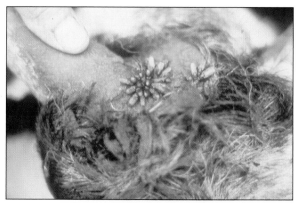

10.6

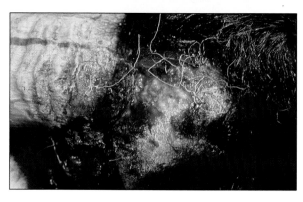

10.59 Headfly strike. Note flies feeding at the base of the horns.

10.60 Headfly strike. Self-mutilation causes oozing of serum and blood, which attracts more flies.

Warble fly infestation

Przhevalskiana silenus infests goats in Mediterranean countries and Asia. *Hypoderma diana* from red deer can infect sheep, but the life-cycle is not completed. Adult flies lay their eggs on hairs of the legs and chest; larvae migrate by a subcutaneous route to the back, where they localise under the cutaneous trunci muscle. The larvae then penetrate the muscle and skin to form breathing holes (**10.61**) and become surrounded by a wall of granulation tissue. Third-stage larvae emerge from the host and fall to the ground to pupate. Damage to hides is severe. Single lesions might be confused with follicular cysts.

10.61

10.61 Warbles. Multiple larvae and their breathing pores can be seen.

Fly strike

Blowfly myiasis occurs when fly larvae invade living tissue. *Lucilia* and *Phormia* spp. are primary flies which can initiate strike while secondary *Calliphora* flies are attracted to already damaged skin.

Soiled fleece, fleece rot, shearing wounds, footrot and headfly lesions all attract flies to lay their eggs (**10.62**). Proteolytic enzymes secreted by blowfly larvae digest tissues. Foul-smelling exudate collects, much skin is underrun or devitalised (**10.63**), and tox-aemia or septicaemia may lead to death.

The screw-worm *Cochliomyia (Callitroga) hominivorax* in North and South America and *Wohlfahrtia magnifica* and *Crysoma bezziana* in the Mediterranean region and Asia are obligate parasites which deposit their eggs at body orifices (**10.64, 10.65**) or in wounds. The larvae are found in clusters and rapidly eat their way into living tissue of sheep and goats as well as other species.

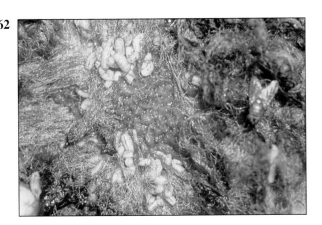

10.62

10.62 Cutaneous myiasis. Numerous maggots are feeding on the skin surface.

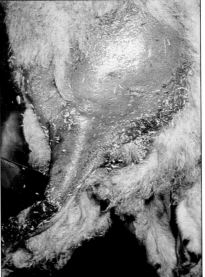

10.63

10.63 Cutaneous myiasis. Feeding fly larvae have devitalised skin on this sheep's leg.

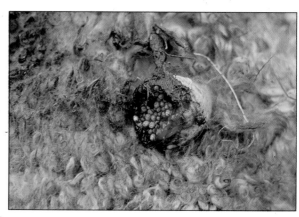

10.64 *Wohlfahrtia* **myiasis.** Clustered larvae have invaded the vulva.

10.65 *Wohlfahrtia* **myiasis.** Larvae are embedded in the skin of the prepuce.

Parelaphostrongylus tenuis skin lesions

Parelaphostrongylus tenuis, the meningeal worm of the white-tailed deer in North America, causes neurological signs when it migrates through the spinal cord or brain of sheep or goats (see Chapter 5). Some goats exhibit one or more linear, vertically orientated skin lesions due to focal self-excoriation (**10.66**). These are presumed to be due to parasitic irritation to the dorsal nerve root supplying sensory innervation to a specific dermatome. Motor nerve deficits of paresis or paralysis are absent in some goats with skin lesions.

Besnoitiosis

Goats and occasionally sheep in Kenya and Iran can be intermediate hosts for tissue cyst-forming protozoa of the genus *Besnoitia*, which has felids as its primary host. Transmission is unknown but may be by biting flies (e.g. *Glossina* spp.). Host macrophages or fibrocytes are parasitised, and become adapted to form large cysts containing bradyzoites in the skin (**10.67**), testis and scleral conjunctiva (see Chapter 8).

Alopecia is associated with white sand-like foci in the subcutis of the neck, limbs and thoracic region. Fibrous reaction around the cysts causes progressive thickening and wrinkling of the skin, with damage to the hides.

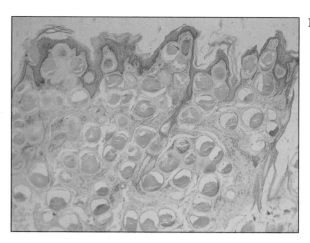

10.66 Parelaphostrongylosis. This paralysed goat has several vertical skin lesions caused by self-excoriation.

10.67 Besnoitiosis. Numerous cysts of *Besnoitia* in the skin of the brisket of a Galla goat. Haematoxylin and eosin.

Immune diseases

Chronic allergic dermatitis

Sheep that are exposed to biting insects on pasture occasionally develop a seasonal, recurrent skin disease which appears to have an allergic component. Unprotected skin, as on the vulva, develops lesions of erythema, oedema and hyperpigmentation (**10.68**). Alopecia and skin damage are increased by self-excoriation in response to pruritus.

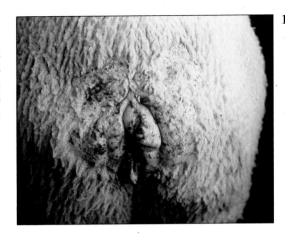

10.68

10.68 Allergic dermatitis. Seasonal pruritus has led to alopecia, erythema and thickening of the skin.

Pemphigus foliaceus

Pemphigus foliaceus is a rare autoimmune disease of goats (also man, dog, cat and horse), where the animal develops autoantibodies against the glycocalyx of its own keratinocytes. Vesicles, pustules and crusts develop over the entire body (**10.69, 10.70**) or are more localised on the perineal region and ventral abdomen. The goat is often pruritic. Diagnosis is by demonstration of intercellular deposition of immunoglobulin by direct immunofluorescence of biopsy specimens. Granular 'cling-ons' (cells from the stratum granulosum) may also be found in direct smears of intact vesicles. Conditions which cause similar lesions (ringworm, see **10.32, 10.33**; staphylococcal dermatitis, see **10.19, 10.20**; or zinc deficiency, see Chapter 15) must be ruled out.

69

10.69 Pemphigus foliaceus. Crusts around the mouth, eyes and ears.

10.70

10.70 Pemphigus foliaceus. Pustules on the inner surface of the pinna.

Nutritional diseases

Copper deficiency

One of the earliest signs of copper deficiency is loss of wool crimp, appearing as an uncrimped band of low tensile strength in the fleece. In breeds with black fleeces, the band may lack pigment. The fleece has a steely texture and lacks tone. These and other aspects of copper deficiency are illustrated in Chapter 15.

Zinc deficiency

Zinc deficiency is often accompanied by hyperkeratosis and crusts, especially on the feet and teats and around the eyes (see Chapter 15).

Seborrhoea sicca

Goats occasionally develop dry scaling or dandruffy skin, especially around the eyes but sometimes involving the entire body (**10.71**). Partial alopecia may also be present. While scales may be secondary to conditions such as ectoparasites, dermatophytes, dermatophilosis, staphylococcal folliculitis or chronic catabolic states, abnormal lipid metabolism and deficiencies of vitamin E or selenium appear to be involved in some goats.

Wool break, wool slip

Individual sheep that are severely stressed by malnutrition or disease undergo an arrest in fibre growth with consequent simultaneous weakening of many wool fibres. When growth resumes and the damaged region nears the skin surface, large masses of wool can be easily pulled off. A similar condition (wool slip) occurs in housed sheep that have been shorn during the winter. Large areas of the body become completely denuded of wool (**10.72, 10.73**). Ectoparasites are absent, and it has been suggested that wool slip results from cold stress, perhaps with over-production of corticosteroids. The differential diagnosis includes ringworm (see **10.32**).

10.72 Wool slip. The sides of this ewe are largely denuded of wool.

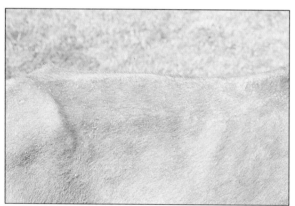

10.71 Seborrhoea sicca. Dandruffy scales and mild alopecia involve much of the hair coat.

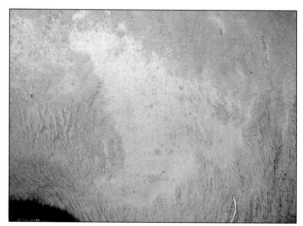

10.73 Wool slip. Alopecia and erythema are evident.

Chemical shearing

Mechanical shearing is expensive and stressful to the sheep, with cuts frequently occurring. Several chemical treatments have therefore been tested as alternatives (**10.74**). Currently, attention is focused on the natural hormone, epidermal growth factor. A single injection induces a break in the wool, and the entire fleece can be peeled off six weeks later.

Cysts and neoplasms

Cysts

Wattle cysts (**1.43**) in goats are usually attached to the base of one or both wattles. Dermoid cysts may occur anywhere on the body but are often found on the back of sheep.

Warts, papillomas

Warts of little clinical significance occur occasionally on the sparsely haired parts of the face or ears of sheep and goats. They are probably caused by papilloma viruses, and exposure to direct sunlight may be a factor in their pathogenesis.

Warts presenting on the white udders of goats are contagious (**10.75**). They may regress spontaneously, persist, or transform into squamous cell carcinomas, which eventually erode through the wall of the teat. Warts at the teat orifice predispose to bacterial mastitis. Other squamous cell carcinomas are illustrated in Chapter 18.

Cutaneous horn

An apparently contagious condition has been seen in sheep originally housed with a goat with udder warts or shorn with equipment used to shear affected sheep. Multiple large horny structures filled with pasty necrotic material protrude from the skin (**10.76**) and persist for years.

10.74

10.74 Chemical shearing. This sheep was treated with cyclophosphamide.

10.75

10.75 Udder warts. Multiple horny warts can be seen on this Saanen doe's udder, while others have been pulled or rubbed off, leaving a raw skin defect.

10.76

10.76 Cutaneous horn. These protruding masses may be 10 cm or more in diameter.

Injuries to the skin

Keloid

A keloid or proliferative scar sometimes forms at the site of a previous skin injury (**10.77**).

Grass or barley awn cellulitis

Plant awns from pasture or bedding can become trapped in the fleece and penetrate the skin of sheep. Bacteria carried in with the foreign matter then induce focal abscesses or more generalised cellulitis (**10.78**).

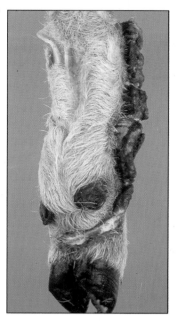

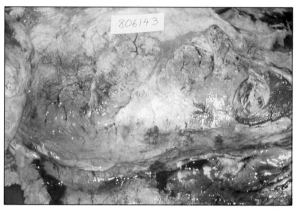

10.77 Keloid. A proliferative scar on the pastern marks the site of a previous wire cut.

10.78 Grass awn cellulitis. Bacteria accompanying migrating plant awns have caused grey-green areas of necrosis.

Frostbite

Lambs and kids born in cold environments may suffer frostbite of the ear tips (or less commonly the feet) if the extremities are not rapidly dried. After the gangrenous area has sloughed, the remaining pinna is of variable length and has a round hairless tip (**10.79**). The condition should be differentiated from squamous cell carcinoma (see Chapter 18) and the normal short ear of the LaMancha goat.

10.79 Frostbite. The ear tips are irregularly shortened and thickened.

Sunburn

This condition is distinct from photosensitisation in that the cause is solely the direct damaging effects of shortwave ultraviolet light on the unprotected skin. It occurs in sheep of the white-faced breeds and in light-coloured goats. The ears and face of sheep and the udders of goats are the sites most commonly affected.

In sheep, erythema is followed by swelling and exudation of serum, with crust formation (**10.80**). Irritation leads to head-shaking, and rubbing leads to loss of hair over the ears and face. Prolonged exposure may cause carcinoma of the pinna (see Chapter 18).

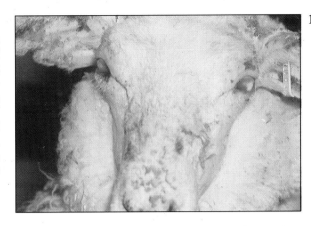

10.80

10.80 Sunburn. The areas with the least protective wool on this sheep's face are red and swollen.

Photosensitisation, yellowses, plochteach

This may be primary—due to consumption of plants containing photodynamic substances, e.g. St. John's wort (*Hypericum* spp.) and some brassicas, for example, rape—or hepatogenous, where liver damage by a whole range of toxic agents results in interference with the excretion of the photodynamic substance phylloerythrin. Phylloerythrin is a normal microbial degradation product of chlorophyll. Presence of phylloerythrin in the unprotected skin (face, ears, limbs or back) causes abnormal sensitivity to direct sunlight, particularly in the white-faced breeds. In the condition alveld, found in Norway, the toxic plant is bog asphodel (*Narthecium ossifragum*).

Affected sheep show acute oedema of the ears and eyelids within a few hours of exposure to sunlight (**10.81**). Animals are often greatly distressed and rub themselves violently to relieve the irritation. Serum oozes from the swollen areas and dries on the face, giving rise to the name 'yellowses'. Later, the affected skin undergoes necrosis and sloughs. Shrivelling and sloughing of the ear tips (**10.82**) may also occur. This is called 'plochteach' in sheep in Scotland, where the cause has not been identified. Necrosis of the ears needs to be differentiated from frostbite (see **10.79**).

.81

10.81 Photosensitisation. The muzzle, eyelids and ears are oedematous.

10.82

10.82 Photosensitisation. Erythema, alopecia and crusting on the muzzle and ear tips. Note the shrivelled ear pinnae.

Facial eczema

This disease is an acute hepatogenous photosensitivity which occurs commonly in sheep in Australia and New Zealand. Goats are relatively resistant. Liver damage is induced by the mycotoxin, sporidesmin, following ingestion of spores of the fungus *Pythomyces chartarum* (**10.83**), found on ryegrass pastures. The fungus proliferates in dead stalks, especially in warm, moist conditions. It is also consumed in association with the poisonous plant *Tribulus terrestris* in South Africa, causing a photosensitising disease, geeldikkop, which closely resembles facial eczema.

After a two-to-three week latent period, there is a very rapid onset of swelling of the face, especially the lips, the dorsal part of the ears and above the eyes (**10.84**). The differential diagnosis includes bluetongue (see Chapter 2) and the intermandibular oedema associated with parasitism. Affected animals seek shade. Later the skin lesions become leathery and hard (**10.85**), and sloughing of the skin may cause death.

Liver failure (see Chapter 3) also contributes to high mortality. The liver in facial eczema is markedly swollen and discoloured (**10.86**). There is necrosis of bile ducts as well as bile duct proliferation, fibroplasia around ducts and obliteration of ducts by scar tissue (**10.87**, **10.88**). An ulcerative cholecystitis is also present.

10.83

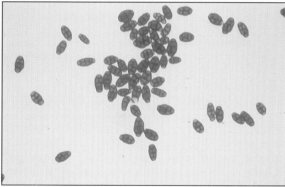

10.83 *Pithomyces chartarum.* Sporidesmin produced by the fungus causes hepatic photosensitisation.

10.

10.84 Facial eczema. Swelling of the muzzle, eyelids and ears in acute disease.

10.85

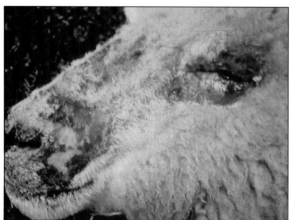

10.85 Facial eczema. Sloughing of skin on the face in chronic disease.

10.

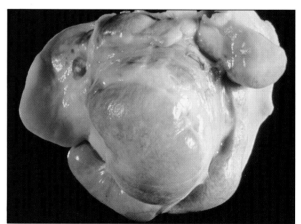

10.86 Facial eczema. This chronically affected liver is pale with peripheral areas of complete atrophy and large, swollen nodules of regeneration.

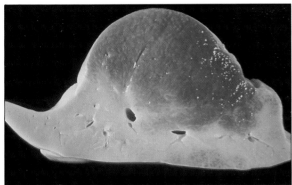

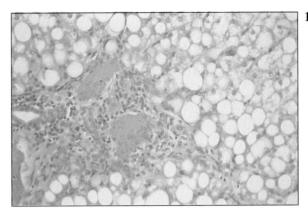

10.87 Facial eczema. Pale tissue on the cut surface of this liver represents relative fibrosis because of atrophy and absolute fibrosis from the effects of toxin. The dark red tissue is regenerating.

10.88 Facial eczema. Lipidosis, bile duct proliferation and chronic inflammatory cell response. Haematoxylin and eosin.

Thermal burns

Severe burns occur to the skin of animals trapped in barn or brush fires. The diagnosis is usually obvious from history and the presence of scorched wool or hair (**10.89**). Skin will slough and scars form after full-thickness burns if the animal survives.

Urine scald

Moist dermatitis with discoloration of hair, alopecia and scabbing develops on parts of the body that are continuously wet with urine. This occurs very commonly on the face and front legs of bucks during the breeding season (**10.90**), as the animals urinate on themselves when sexually aroused. Urine scald also occurs on the escutcheon or inner thigh of intersex animals with abnormal external genitalia, and of male animals after urethrostomy. Urine can contribute to a ventral contact dermatitis in animals housed under very unsanitary conditions.

10.89 Thermal burns. Full-thickness skin burns on the legs and scorched wool are evident on this sheep that died in a brush fire.

10.90 Urine scald. Lesions on the face and lips occur when the buck urinates on himself.

Oiling

Sheep are occasionally exposed to and become covered with crude oil (**10.91**) when they graze along a coast where oil spills have washed ashore.

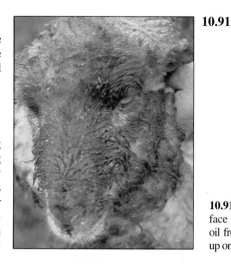

10.91

10.91 Oiling. This sheep's face is covered with crude oil from an oil slick washed up on the shore.

Dog worry, bitten ears

Dogs frequently attack sheep and goats, killing some by severing the jugular veins or inducing pulmonary oedema. Inexperienced dogs may cause multiple wounds to the skin or joints (**10.92**). Attacks on sheep and goats by other domestic species, including cattle, are not uncommon when mixed grazing is allowed. Sheep have been killed deliberately or had their ears bitten by aggressive horses (**10.93**).

10.92

10.92 Bite wounds. This goat was tethered when attacked by dogs.

10.

10.93 Bite wounds. This sheep's ears were cropped by a horse.

Lynx bites

Wild cats occasionally attack and kill sheep or goats. Bite wounds should be searched for on the ventral surface of the neck (**10.94**).

Buckshot wounds

Man is also a dangerous predator for sheep and goats. Gunshot wounds in the skin are easily overlooked if the undersurface is not examined (**10.95**).

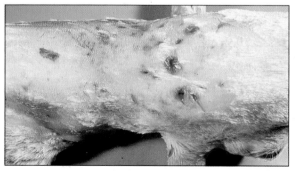

10.94

10.94 Lynx bites. Clipping hair off the neck of this goat has exposed puncture wounds inflicted by a lynx.

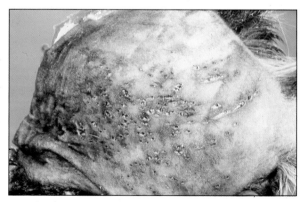

10.

10.95 Gunshot wounds. The many holes visible on the underside of the skin were caused by buckshot.

11 Diseases of the Urinary System

In this chapter are illustrated a number of common infectious diseases involving the kidneys and lower urinary tract. Additionally, examples of obstructive diseases of the urinary tract are shown. For specific toxic diseases of the kidneys, the reader is referred to Chapter 17, while neoplasms of the kidneys and adrenals are described in Chapter 18.

Diseases of the kidneys

Amyloidosis
Deposition of amyloid can occur in glomeruli and in the renal medulla of small ruminants in cases of idiopathic generalised amyloidosis (see Chapter 4). Amyloid is deposited in the glomerular mesangium, subendothelially in the glomerular capillaries and along the basement membranes of the tubules. It is demonstrated by polarised light after Congo red staining (**11.1**).

Caseous lymphadenitis
The kidney is an occasional site for chronic abscesses (**11.2, 11.3**) in disseminated *Corynebacterium pseudotuberculosis* infections of sheep and goats (see Chapter 14). The frequently laminated appearance of the lesions distinguishes them from those of melioidosis (see page 180).

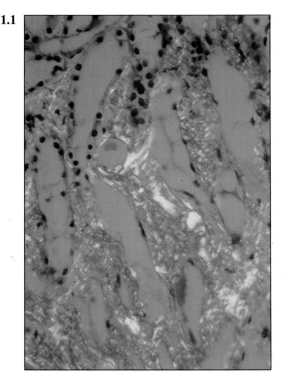

11.1 Renal amyloidosis. Birefringence of amyloid deposits in polarised light after Congo red staining.

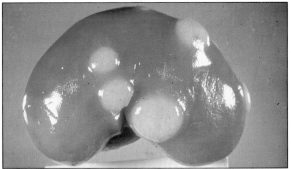

11.2 Caseous lymphadenitis. Abscesses in a sheep kidney.

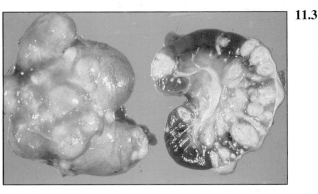

11.3 Caseous lymphadenitis. A kidney sectioned to show laminated lesions.

Chlamydial nephritis

Nephritis is an unusual complication of the chlamy-daemic phase of infection with *Chlamydia psittaci* in sheep. Necrotic foci are found in the kidneys at necropsy (**11.4**).

11.4

11.4 Chlamydial nephritis. Focal necrotic lesions in a sheep kidney.

Clostridial/coliform necrosis

Necrotic lesions, from which both *Escherichia coli* and *Clostridium perfringens* (types C or D) are isolated, are occasionally found in sheep at necropsy (**11.5**). The pathogenesis of these dual infections is not clear.

11.5

11.5 *Escherichia coli* and *Clostridium perfringens* infection in a sheep kidney.

Diabetes insipidus

Natural ocurrences of diabetes insipidus occasionally occur in sheep. Enlargement of the kidneys with bulging of the cut surface, friability and widening of the renal pelvis are the main macroscopic features in this seven-month-old Suffolk-cross lamb which had polydipsia with polyuria (**11.6, 11.7**).

On microscopic examination, the capsular lining cells and the cortical tubular epithelium contain haemosiderin pigment, while focal glomerular atrophy, hyalinisation of Bowman's capsules and fibrosis are also present. The fundamental brain lesion is absence of the posterior pituitary and infundibular tissue.

11.6

11.6 Diabetes insipidus. Abnormal shape of an affected sheep kidney.

11.7

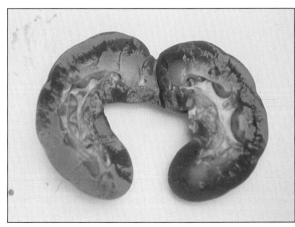

11.7 Diabetes insipidus. A kidney opened to show the narrow cortex and dilated renal pelvis.

Infarction

The most common cause of renal infarcts is embolism from vegetations on the heart valves (see Chapter 13). Wedge-shaped necrotic foci extend through the entire depth of the cortex (**11.8**).

Leptospirosis

Leptospires of the *Leptospira interrogans* serogroups *pomona, australis, ballum, hardjo, sejroe* and *grippotyphosa* have been implicated in clinical disease in sheep, while serogroups *grippotyphosa, icterohaemorrhagiae, pomona* and *sejroe* can cause disease in goats. *L. pomona* and *L. grippotyphosa* serovars produce haemolysins which cause haemoglobinuria and acute icterus. *L. hardjo,* probably the most prevalent serovar, mainly causes chronic interstitial nephritis. All serogroups can cause sporadic abortions and stillbirths, and serovar *hardjo* can cause agalactia in ewes.

In acute cases, the carcase is icteric and anaemic (**11.9**), the liver friable with focal areas of necrosis, and the kidneys swollen and dark due to the presence of haemoglobin breakdown products in the tubules (**11.10**). Diagnosis is usually made by dark-field microscopy of urine or blood, or demonstration of the organisms in renal tissues processed by the silver staining methods of Levaditi or Warthin–Starry.

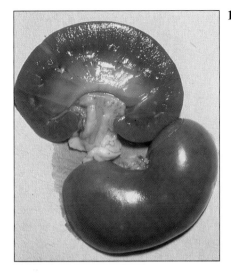

11.8

11.8 Renal infarcts in staphylococcal endocarditis. Note the pale wedge-shaped necrotic area in the cortex.

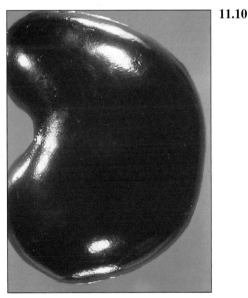

11.10

11.10 *Leptospira pomona* infection. Discoloured kidney due to excretion of haemoglobin products.

11.9

11.9 *Leptospira pomona* infection. Icteric carcase of an infected lamb.

11.11

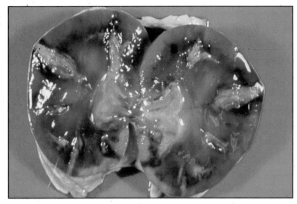

11.11 Melioidosis. Abscesses in the kidney.

Melioidosis

This chronic infection of sheep or goats with *Pseudomonas pseudomallei* can give rise to multiple abscesses in the lymph glands and other organs, including the kidneys **(11.11)**. The abscesses can be confused with those found in caseous lymphadenitis.

Membranous glomerulonephritis

Occasionally, pale contracted kidneys are found at necropsy in animals with chronic abscesses or other long-standing infections. Microscopic examination reveals diffuse glomerulosclerosis with narrowing or occlusion of capillaries **(11.12)**. Silver-staining methods show that glomerular capillaries have a 'cog-wheel' appearance due to the presence of subepithelial dense deposits and basement membrane thickening **(11.13)**.

11.12

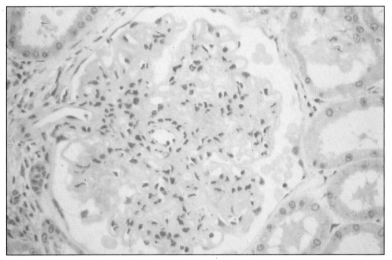

11.12 Membranous glomerulonephritis. Ischaemia and fibrosis of a glomerulus. Haematoxylin and eosin

11.13

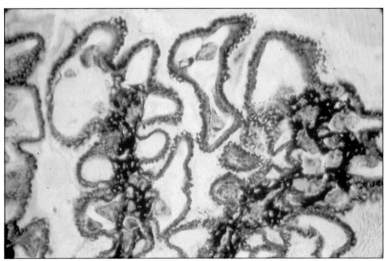

11.13 Membranous glomerulonephritis. 'Cog-wheel' effect of silver-stained subepithelial dense deposits in capillary loops.

Mesangiocapillary glomerulonephritis

This is an hereditary disease of Finnish Landrace sheep or their crosses. Fundamentally an immune-complex glomerulonephritis, the basic cause appears to be a genetically controlled deficiency of the third component of the complement cascade.

Affected lambs are clinically normal at birth, though glomerular changes can be detected microscopically. Within a few weeks, the kidneys are palpably enlarged and tender, and a variety of clinical signs including inappetance, circling, nystagmus and facial twitching lead progressively to convulsions and death from renal failure.

At necropsy, the kidneys are enlarged and 'flea-bitten' with red specks (**11.14**) which are haemorrhagic glomeruli. Some chronically affected lambs may have large kidneys with yellowish, slightly translucent cortices in which the fibrosed glomeruli stand out as white spots (**11.15**). The glomerular changes include mesangial hypercellularity and capillary wall thickening (**11.16**), due to subendothelial deposition of immune complexes. Focal haemorrhage and oedema may be found in the cerebrum (see Chapter 5).

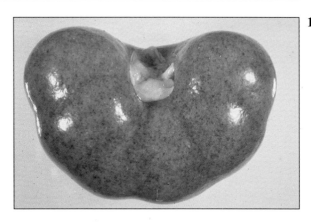

11.14

11.14 Mesangiocapillary glomerulonephritis. Flea-bitten kidney.

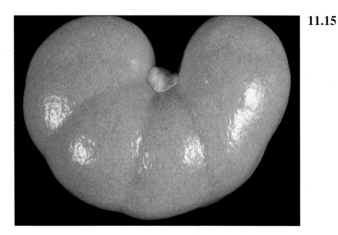

11.15

11.15 Mesangiocapillary glomerulonephritis. Chronic renal changes. The white flecks are sclerosed glomeruli.

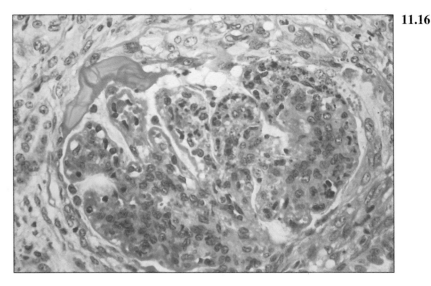

11.16

11.16 Mesangiocapillary glomerulo-nephritis. A glomerulus showing mesangial proliferation, capillary wall thickening, and fibrin in Bowman's capsule. Haematoxylin and eosin.

Nephrosis

This term has been used in the United Kingdom to describe a syndrome in lambs between two and 12 weeks old at pasture in the late spring and early summer. In this condition, young lambs up to about four weeks old cease sucking and become lethargic and incoordinated (drunken lamb syndrome) before dying of renal failure after a few days. Older lambs are bright initially but often have diarrhoea. They do not respond to anthelmintic or other therapy and soon pine and die.

At necropsy, both kidneys are often greatly enlarged and very pale (**11.17**). On microscopic examination, the proximal convoluted tubules are dilated and contain either hyaline material or coarse granular material with the staining characteristics of fibrin. In older lambs there is an increase in cortical stroma. The disease is non-inflammatory and the evidence of pathological and biochemical studies suggests a nephrotoxic cause.

11.17

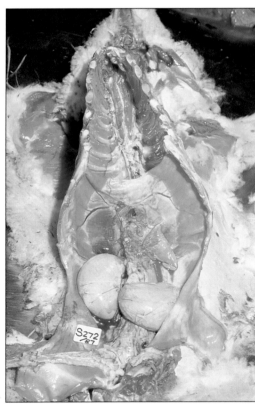

11.17 Nephrosis. A lamb with typical swollen, pale kidneys.

Non-specific tubular necrosis

Non-specific toxic tubular necrosis is a common finding in sheep at necropsy. The kidneys are swollen, tan-coloured and have a watery appearance. Necrotic groups of tubules produce a pale, stippled effect under the capsule (**11.18**). The cause of the necrosis is seldom identified.

11.1

11.18 Toxic tubular necrosis. Macroscopic appearance of the kidney.

Pyaemic nephritis

Pyogenic bacteria carried in the bloodstream can become trapped in renal capillaries (**11.19**), where they multiply rapidly and induce local necrosis and abscess formation. In peracute infections, the kidneys are enlarged and have dark foci in the cortex. The presence of microabscesses can be confirmed microscopically.

11.1

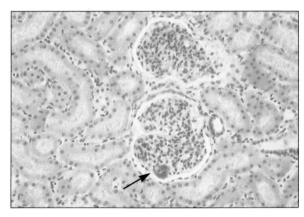

11.19 Pyaemic nephritis. Bacterial embolus trapped in glomerular capillary (arrow): a nidus for abscess formation. Haematoxylin and eosin.

Pyelonephritis

Ascending infections from the lower urinary tract (**11.20**) can lead to acute (**11.21**), or chronic infection of the renal pelvis. These infections are often derived from contamination by intestinal flora, and are more common in females. Local extension of infection into the medulla can occur. Wedge-shaped areas of cortex may contain dilated tubules that have become obstructed, and which contain purulent or hyaline casts. These areas eventually become atrophic and fibrotic. Where the infection is caused by pyogenic bacteria, suppurative pyelitis and pyelonephritis (**11.22**) are the outcome.

11.20 Pyelonephritis. Cystitis and suppurative urethritis predisposing to infection of the renal pelvis.

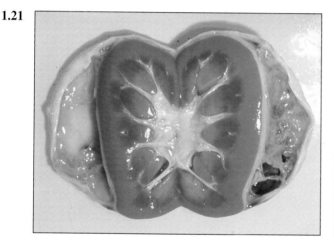

11.21 Pyelonephritis. Acute subcapsular oedema in acute infection.

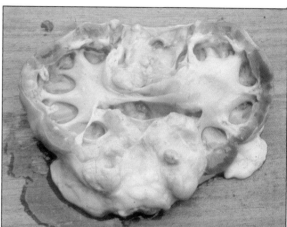

11.22 Pyelonephritis. Chronic suppuration in a sheep kidney pelvis.

Disease of the urinary bladder

Cystic calculi

Large calculi of the type shown in **11.23** are quite commonly found in the bladders of clinically normal sheep at slaughter.

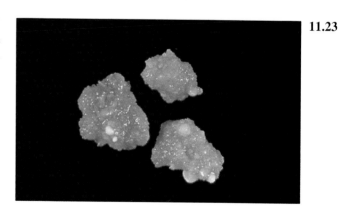

11.23 Typical cystic calculi.

11.24

11.24 Acute cystitis in a wether following urethral obstruction.

Cystitis

Inflammation of the bladder (**11.24**) almost invariably is a consequence of the chronic distension and devitalisation which occur in urolithiasis (see page 185). The organisms commonly isolated include *Escherichia coli, Proteus vulgaris*, streptococci and staphylococci.

Obstructive diseases of the urinary tract

Renal calculi

Renal calculi, which are fairly common findings in sheep at necropsy (**11.25**), are usually mixtures of calcium, magnesium or ammonium phosphates in a mucoprotein matrix. Problems arise when aggregates of calculi cause blockages in the ureter (see below). If this is unilateral, it may go undetected.

Rubber castration ring causing urethral obstruction

Inappropriate positioning of the rubber ring used for castration caused urethral obstruction in the lamb shown in **11.26.**

11.25

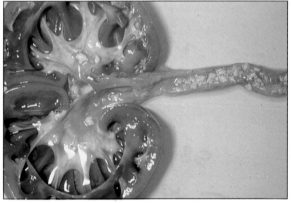

11.25 Renal calculi. The renal pelvis is dilated due to obstruction of the ureter.

11.26

11.26 Urethral obstruction by a rubber castration ring.

Ureteral obstruction

Renal calculi sometimes pass into and obstruct one or, occasionally, both ureters. The usual necropsy finding is unilateral enlargement of the kidney (**11.27**).

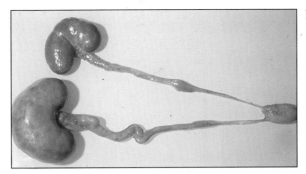

11.27 Ureteral obstruction. Unilateral enlargement of the kidney.

Urolithiasis

This is a sequel of urinary calculus formation, and occurs in male lambs, kids or wethers fed on cereal-based concentrate rations. Obstruction can occur at the sigmoid flexure and along the body of the urethra, though the most common manifestation is the presence of a calculus just proximal to, or within, the urethral process (**11.28**).

Massive distension and congestion of the bladder occur, and damming back of urine causes swelling of the kidneys. Affected animals are distressed and strain violently in an attempt to urinate. Crystalline calculi may be found clinging to the preputial hairs. Rupture of the urinary tract can occur at any point, although the bladder often ruptures first. This relieves the distress in the short term, but toxaemia quickly supervenes. Affected animals may be found recumbent, with pronounced ventral oedema (**11.29**).

At necropsy, the bladder may be distended and inflamed, or ruptured, in which case the abdominal cavity will be filled with urine. The kidneys may be swollen, dark and surrounded by oedematous fat. Calculi may be found at all levels of the urinary tract, including the renal pelvis, ureters, bladder and urethra.

In sheep, calculi are composed commonly of struvite (magnesium ammonium phosphate hexahydrate), silica, oxalates or carbonates. Sheep grazing on oestrogenic pastures in Australia can develop 'clover stones' (benzocoumarins). Less commonly found in sheep are

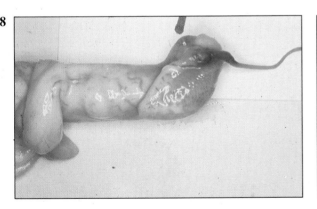

11.28 Urolithiasis. Calculus in the processus urethrae of a wether.

11.29 Urolithiasis. Ventral oedema in an affected sheep due to urethral rupture.

11.30

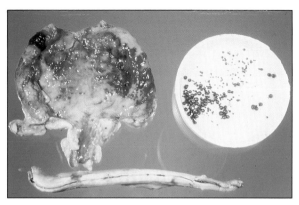

11.30 Urolithiasis. Calcium carbonate calculi in a goat penis, with associated cystitis.

xanthine calculi. The commonest calculus in goats is struvite, but gold-coloured spherical calculi composed of calcium carbonate **(11.30)** are occasionally seen. Diet, urinary pH and availability of water supplies influence their formation.

12 Diseases of the Eye

Eye diseases, which are often painful and debilitating, are important not just for reasons of economic loss but for considerations of welfare. This chapter illustrates a number of common ocular conditions of sheep and goats caused by bacteria and mycoplasmas. Certain toxic plants can cause retinal degeneration, and examples of this lesion are included. Miscellaneous conditions include the effects of migrating parasitic larvae and a comment on entropion.

The ocular lesions are grouped according to the structures involved. The major categories are infectious or traumatic keratoconjunctivitis, anterior uveitis and retinal degeneration. When no ocular lesion is detected but the animal is partially or completely blind, consideration should be given to central nervous system diseases, such as cerebrocortical necrosis or coenurosis (see Chapter 5). Vitamin A deficiency can be a cause of night blindness in arid environments.

Keratoconjunctivitis

Entropion

In the neonatal lamb or kid, a congenital entropion (usually involving the lower eyelid) is a common cause of traumatic keratitis. Epiphora and squinting are noted from a distance (12.1) but are also typical of pink eye. The lower lid rolls inwards until hairs rub on the surface of the cornea (12.2). Neovascularisation of the cornea is commonly present. The cornea heals quickly after the eyelid position is corrected. Acquired entropion occurs in adult animals secondary to painful ocular conditions (spastic entropion) or sinking of the globe caused by severe weight loss or dehydration.

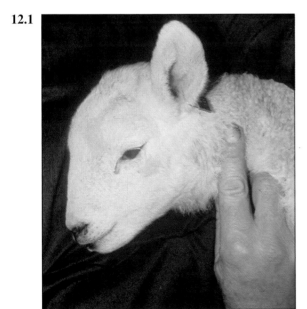

12.1 Entropion. The affected eye is held closed.

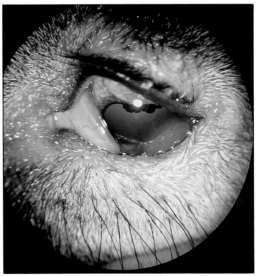

12.2 Congenital entropion. The lower lid margin is not visible, and granulation tissue is present where hairs have abraded the cornea.

Infectious keratoconjunctivitis

Keratoconjunctivitis (pink eye) often occurs as an infectious flock or herd problem. The aetiology cannot be determined without extensive laboratory back-up. *Mycoplasma conjunctivae, M. agalactiae, Chlamydia psittaci* (polyarthritis immunobiotype) and bacteria have all been incriminated. The severity of infection is increased when more than one agent is involved. *M. conjunctivae* remains in the eye long after clinical signs disappear, and carrier animals are commonly a source of infection to other herds and flocks.

The clinical signs are those of conjunctivitis with or without keratitis. Lachrimation and blepharospasm are visible from a distance (**12.3**). There is mild-to-severe hyperaemia of the conjunctiva, and lymphoid follicles may develop (**12.4**), especially with chlamydial infection. A mucopurulent discharge commonly collects in the conjunctival sac and on the surface of the cornea.

With corneal involvement, there is initially a peripheral neovascularisation (**12.5**). Pannus formation and oedema may render the entire cornea opaque and the animal temporarily blind (**12.6,12.7**). Central ulceration does not usually occur unless there is bacterial infection, as with *Branhamella (Neisseria) ovis*. When only a central opacity is present, *B. ovis* is usually the only organism that can be demonstrated (**12.8**).

Conjunctival cultures and scrapings are used to identify the aetiological agents involved in keratoconjunctivitis. Interpretation of Wright or Giemsa-stained smears (**12.9**) is complicated by the presence of melanin granules or stain precipitate. Fluorescent antibody tests on smears made from scrapings can be used to specifically identify mycoplasmas or chlamydia (**12.10**).

12.3

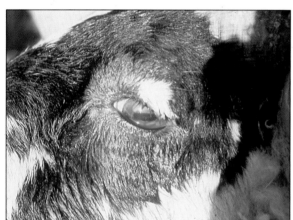

12.3 Infectious keratoconjunctivitis. Lachrimation is evident on the face of this Scottish Blackface.

12.5

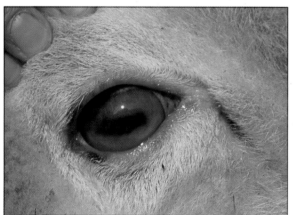

12.5 Infectious keratoconjunctivitis. Peripheral neovascularisation induced by *Mycoplasma conjunctivae*.

12.4

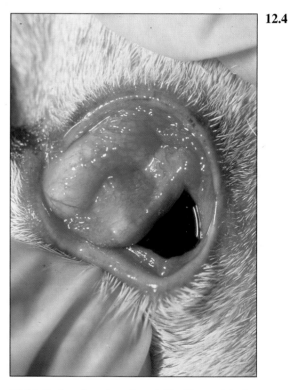

12.4 Infectious keratoconjunctivitis. The conjunctiva is reddened and swollen with marked follicle formation.

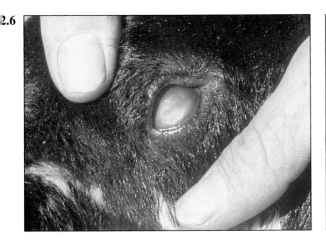

12.6 Infectious keratoconjunctivitis. An advanced lesion in which the cornea is oedematous and opaque.

12.7 Infectious keratitis. Advanced keratitis caused by *Mycoplasma agalactiae*.

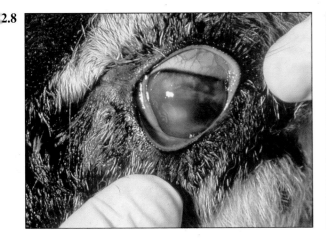

12.8 Infectious keratitis. Central corneal opacity in the absence of other ocular changes is usually associated with *Branhamella ovis*.

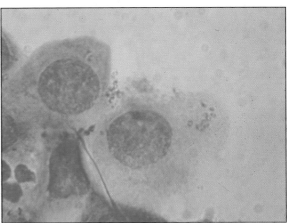

12.9 Infectious keratoconjunctivitis. Conjunctival scraping from a mature ewe with keratoconjunctivitis. Mycoplasmas are visible in the cytoplasm of epithelial cells. Giemsa stain.

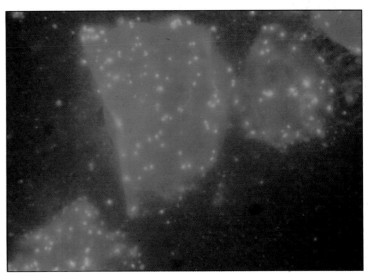

12.10 Infectious keratconjunctivitis. *Mycoplasma conjunctivae* is identified by an indirect fluorescent antibody test.

12.11

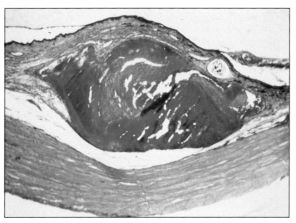

12.11 *Gedoelstia hassleri.* The bot fly larva within a blood vessel has caused thrombosis. Haematoxylin and eosin.

Parasitic keratoconjunctivitis

Besnoitia besnoiti, a protozoan parasite, occurs in sand-like ocular cysts on the scleral conjunctiva of goats. The same parasite causes dermatitis and infertility (see Chapters 8 and 10).

The nematode eyeworm, *Thelazia californiensis*, is sometimes found in the conjunctival sac or on the surface of the cornea. It causes discomfort and epiphora. In South Africa, migrating larvae of the nasal bot flies *Gedoelstia hassleri* and *G. cristata* cause keratoconjunctivitis and panophthalmitis. Vasculitis and thrombosis (**12.11**) lead to oedema, haemorrhage and a bulging eye. Cataracts are common. Sheep are more frequently affected than goats. Trypanosomes and screw-worms can also cause keratoconjunctivitis.

Involvement of deeper ocular structures

12.12

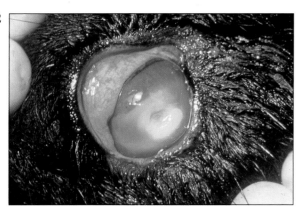

12.12 Hypopyon. Note the keratitis (peripheral neovascularisation and central ulcer) and accumulation of pus in the anterior chamber.

Septicaemia (as from omphalophlebitis or systemic mycoplasma infection) can lead to anterior uveitis and an accumulation of fibrin or pus in the anterior chamber. Hypopyon also occurs after perforation of a corneal ulcer or traumatic injury (**12.12**). Healing often leaves the eye comfortable but permanently blind (**12.13**). Listeriosis is another condition in which both keratitis (from exposure) and an anterior uveitis may be present (**12.14**).

12.13

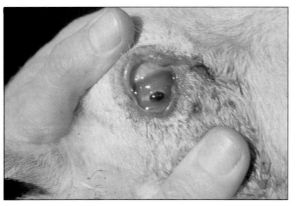

12.13 Ruptured eyeball. This shrunken eyeball with central iris prolapse was the result of a traumatic or infectious perforation of the cornea.

12.

12.14 Exposure keratitis. This Nubian goat with facial nerve paralysis has a total opacity of the cornea. Both drying of the cornea and anterior uveitis induced by *Listeria* could be involved.

Bright blindness

Although clinicians rarely examine the fundus of small ruminant eyes (**12.15, 12.16**), several conditions interfering with vision affect only the posterior segment of the eye.

Ingestion of bracken fern (*Pteridium aquilinum*) by sheep over a long period of time, in addition to oncogenic effects (see Chapter 18), causes progressive retinal atrophy and blindness in animals over two years of age. This is associated with marked narrowing of the retinal arteries and veins (**12.17, 12.18**). The reflectivity of the tapetum to light is altered so that the eyes shine abnormally brightly in semi-darkness (**12.19**). Histological examination reveals degeneration of the neuroepithelium of the retina (**12.20, 12.21**) with destruction of rods, cones and the nuclear layer.

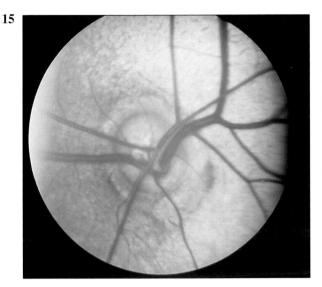

12.15 Goat fundus. Normal Nubian goat.

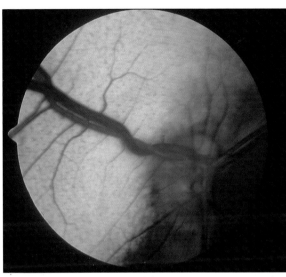

12.16 Sheep fundus. Normal tapetum, blood vessels and optic disc are visible.

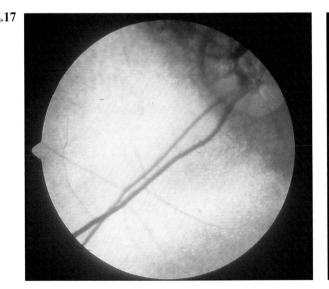

12.17 Bright blindness. The retinal arteries are narrowed. Compare with **12.18.**

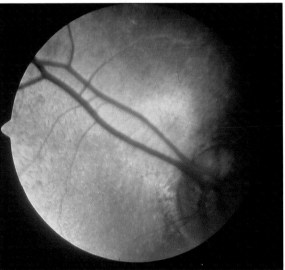

12.18 Normal sheep fundus showing the calibre of the normal arteries.

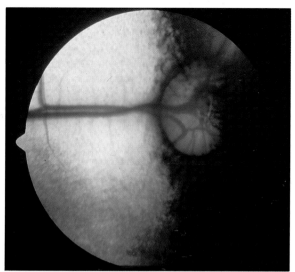

12.19 Bright blindness. The tapetum is hyper-reflective.

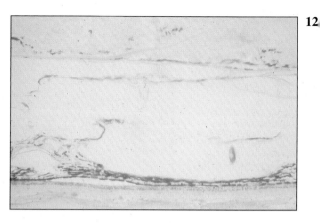

12.20 Bright blindness. Severe retinal atrophy. Haematoxylin and eosin.

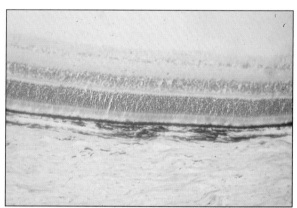

12.21 A normal sheep retina. Haematoxylin and eosin.

Stypandra blindness

Stypandra imbricata (blindgrass) and *S. glauca* (nodding blue lily) are plants that grow in western Australia. Sheep and goats become permanently blind, with no pupillary reflexes, if they survive the acute toxicity associated with eating these plants. There is degeneration of photoreceptor cells, optic nerves and optic tracts (**12.22**). Focal pigment epithelium hypertrophy develops in the nontapetal fundus several weeks after consumption of *S. imbricata*.

12.22 Stypandra blindness. Poisoning by blindgrass *(Stypandra glauca).* Complete disorganisation of layers of the retina, absence of photoreceptor cells and hypertrophy of pigment epithelial cells (large, round dark bodies). Haematoxylin and eosin.

13 Diseases of the Cardiovascular System

The heart and its outer sheath, the pericardium, are vulnerable to infection in a number of common septicaemic diseases of sheep and goats. The myocardium can be the site for migrating nematodes or *Sarcocystis* spp., and for tapeworm cysts. The myocardium is also susceptible to the effects of certain metabolic diseases.

Specific diseases of blood vessels are fortunately rare in small ruminants, although several are illustrated. Cardiac defects are dealt with in Chapter 1, while toxic conditions involving the heart are illustrated in Chapter 17. The tropical disease heartwater is illustrated in Chapter 16.

Diseases that affect the pericardium

Fibrinous pericarditis

The most common cause of fibrinous pericarditis is the septicaemic form of *Pasteurella haemolytica* biotype A infections, although in lambs *Streptococcus* spp. septicaemia can cause a similar picture.

A thick coating of coagulated fibrin covers the pericardium and epicardium, giving a shaggy appearance when the pericardium is reflected; this is called bread-and-butter pericarditis (**13.1**)

Staphylococcal pericarditis

This condition is caused by *Staphylococcus aureus*. An acute inflammation results from infection (**13.2**), and a sero-haemorrhagic and foul-smelling exudate which may contain strands of floating fibrin is produced. Subepicardial petechiae are generally also present.

13.1 Fibrinous pericarditis. Bread-and-butter lesions in a goat.

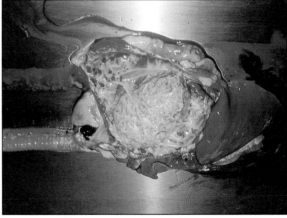

13.2 Staphylococcal pericarditis. Fibrinous reaction in *Staphylococcus aureus* infection.

13.3

Traumatic pericarditis

Sharp foreign bodies in the reticulum occasionally penetrate the diaphragm and the pericardium (**13.3**), where they set up an acute and often fatal pericarditis.

13.3 Traumatic pericarditis in a goat caused by a hypodermic needle.

Diseases that affect the myocardium

Cerebrocortical necrosis heart lesions

An occasional complication of cerebrocortical necrosis in sheep and goats is acute myocardial degeneration (**13.4**). This must be differentiated from nutritional myopathy (see Chapter 15).

Cysticercus ovis

Cysts of the sheep tapeworm *Taenia ovis* are sometimes found in the myocardium at slaughter (**13.5**).

13.4

13.4 Myocardial degeneration in cerebrocortical necrosis.

13.

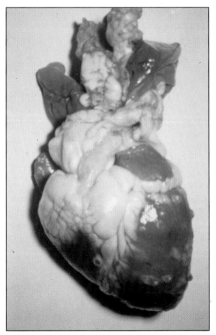

13.5 *Cysticercus ovis* cysts in the heart.

Nodular worm abscess

The heart is an occasional site for migrant larvae of the nodular worm *Oesophagostomum columbianum*. Encapsulated granulomatous or mineralised lesions form around the degenerating parasite (**13.6**), which fails to mature in this site.

13.6 Nodular worm abscess in a sheep heart.

Sarcocystosis

The myocardium is a common site for mature cysts of *Sarcocystis tenella* (*S. ovicanis*). Although they may be very numerous, the cysts seldom provoke any inflammatory reaction (**13.7**).

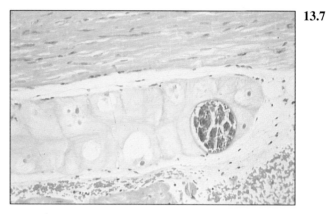

13.7 Sarcocyst in the bundle of His in a sheep. Haematoxylin and eosin.

Diseases that affect the endocardium

Acute endocarditis

Endocarditis can result from acute infections: for example, with enteric *Streptococcus* spp. in lambs and with *Staphylococcus aureus* or *Actinomyces pyogenes* in lambs and sheep. Congenital malformations can predispose to endocarditis.

Acute endocarditis is associated with petechial or ecchymotic haemorrhages beneath the endocardium, as well as with congestion and inflammation (**13.8**).

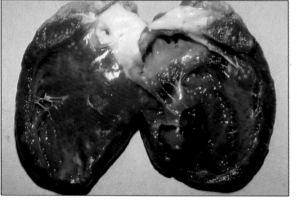

13.8 Acute bacterial endocarditis. Necrotising lesions in a sheep heart.

Cysts on heart valves

These cysts, which evolve from haematomas in the clefts of the fetal valves or from lymphatic channels in the cusps (**13.9**), are harmless unless large enough to cause valvular incompetence.

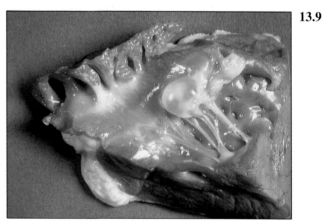

13.9 Cysts on heart valves.

13.10

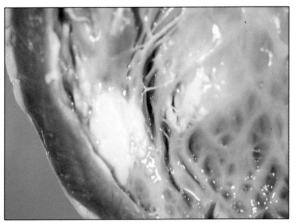

13.10 Endocardial steatosis in a sheep heart.

Endocardial steatosis

Plaques of semi-mineralised fat of no clinical significance and of unknown origin can be found occasionally at slaughter in old sheep or goats (**13.10**).

13.11

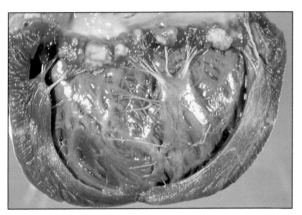

13.11 Verrucose endocarditis. Vegetations on heart valves

Verrucose endocarditis

The heart valves can be a predilection site for bacterial adherence and multiplication; for example, in joint-ill infections, infections with α-haemolytic streptococci or as a sequel to *Erysipelothrix rhusiopathiae* infection.

Large yellowish-red or yellowish-grey vegetations form on affected valves (**13.11**). They are similar in structure to thrombi, and contain colonies of bacteria in pure populations, so that culture will establish an accurate diagnosis. Valvular endocarditis is usually fatal. Shedding of fragments of vegetations causes embolism with infarction in various organs and spread of infection to distant sites. Stenosis or incompetence can result in heart failure, associated with severe passive venous congestion—nutmeg liver (**13.12**).

13.12

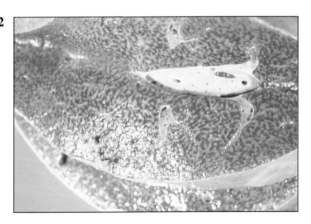

13.12 Nutmeg liver in vegetative endocarditis.

Diseases that affect the blood vessels

Enzootic calcinosis

Several plant species, such as *Solanum malacoxylon*, *Cestrum diurnum* and *Trisetum flavescens,* contain 1,25-dihydroxycholecalciferol and ingestion by animals can cause changes similar to those of hypervitaminosis D.

Soft-tissue calcification occurs, with a predilection for fibroelastic tissues, such as the aorta **(13.13)** and major arteries, but also for alveolar septa, the kidney and the abomasum. The metaphyses of long bones may be obliterated by a fibrillary matrix produced by osteoblasts after intense osteoclastic activity has removed the primary spongiosa by resorption.

13.13 Enzootic calcinosis. Calcification of a sheep aorta due to ingestion of *Trisetum flavescens.*

Intrathoracic haemorrhage

Traumatic rupture of the anterior vena cava can occur **(13.14)**, usually as the result of road or other accidents, but occasionally in rams due to fighting.

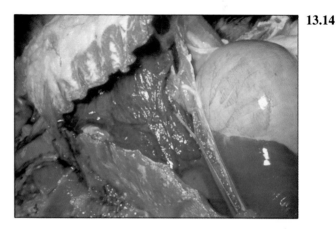

13.14 Rupture of the anterior vena cava with intrathoracic haemorrhage.

Mycotic thrombosis and phlebitis

Intravascular mycosis is uncommon in sheep and goats, although in sheep experimentally immunosuppressed by inoculation of blood containing the tick-borne fever agent *Cytoecetes phagocytophila,* generalised zygomycosis (phycomycosis) is a complicating factor. Should they gain entry to blood vessels, opportunistic fungi flourish **(13.15)**, probably encouraged by the high sugar content of the blood. Fatal mycotic phlebitis with thrombosis is the usual outcome.

13.15 Mycotic phlebitis. Fungal thrombus in the vena azygos of a sheep.

13.16

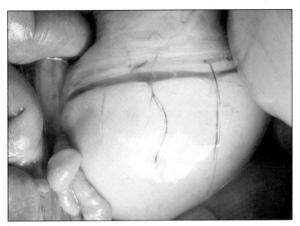

13.16 Schistosomiasis. Schistosomes in the mesenteric veins.

13.17

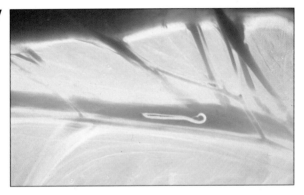

13.17 Schistosomiasis. *Schistosoma bovis* schistosome in a sheep mesenteric vein.

13.18

13.18 Schistosomiasis. Nodular lesions caused by *Schistosoma mattheei* in a sheep liver .

Schistosomiasis

Schistosomes or blood flukes are trematode parasites that live in blood vessels. Several species of *Schistosoma* act as parasites of sheep and goats, depending on geographical location, including *S. mattheei* in Africa, *S. indicum* in India and *S. japonicum* in the Far East. *Schistosoma bovis*, the species which infects mainly bovid species, can also infect sheep and goats. The life cycle of schistosomes is similar to that of most other flukes, except that the redia stage in the intermediate mollusc is replaced by a generation of daughter sporocysts.

Adult schistosomes lay their eggs in host intestinal blood vessels. Each egg containing a miracidium passes through the vessel wall and the intestinal mucosa to escape into the environment. The eggs hatch in water and the miracidia penetrate the soft tissues of aquatic molluscs. Later, infective cercariae emerge from the mollusc and enter the definitive host through either the skin or the alimentary tract. The next stage, or schistosomule, passes to the lungs and heart and then to the veins of the abdominal cavity, particularly the portal and mesenteric vessels (**13.16, 13.17**), where it grows to adulthood. The mature fluke measures up to 3 cm in length and feeds on host red cells.

Animals with schistosomiasis are anaemic and may be severely emaciated. Hypoalbuminaemia can lead to fluid accumulation in serous cavities. The intestinal wall may be thickened due to damage incurred when the spinous eggs are extruded, and granulomatous nodules develop in the intestinal wall, mesentery, regional lymph nodes and liver in heavy infections (**13.18**). Adult schistosomes cause local phlebitis and thrombosis.

14 Diseases of the Haemic and Lymphatic Systems

This chapter deals with a number of diseases which individually signal their existence by overt and excessive destruction or impaired production of circulating erythrocytes, leukocytes or platelets. Certain diseases which specifically localise in the lymphatic system, such as caseous lymphadenitis, or which spontaneously arise as neoplastic processes within the lymphatic system, are also included.

Diseases of the haemic system

Anaemia and icterus

These are not diseases in themselves but are important clinical indicators of a whole range of conditions.

Anaemia

Anaemia occurs as a result of decreased production, increased destruction or loss of erythrocytes. Decreased production can be found in deficiency states such as copper deficiency and perhaps also in cobalt deficiency (see Chapter 15). Chronic parasitic diseases, including fascioliasis, parasitic gastroenteritis or heavy ectoparasite infections, can cause exhaustion of the haemopoietic capacity of the bone marrow (see Chapters 3 and 10).

Haemorrhagic anaemia with rapid loss of red cells is a common sequel to acute parasitism where there is substantial laceration by the parasites; for example, in acute haemonchosis, fascioliasis or hookworm infections (see Chapter 3). Traumatic injury with haemorrhage can result from accidents, surgery or obstetrical trauma, or can occur in lambs or kids as a result of umbilical bleeding. The key clinical sign is extreme pallor of mucous membranes, which is readily seen by examination of the conjunctivae (**14.1**). The pallor is caused by deficiency of haemoglobin from loss of red cells; this is readily confirmed by reference to a haematocrit, which will show a reduced packed cell volume. Oedema may occur in dependent tissues of the body, such as the submandibular space (bottle jaw).

Necropsy findings in non-haemolytic anaemias include fatty change in liver and kidneys, flabby, watery heart muscle, and fluid in body cavities. The spleen may be contracted. In chronic anaemias, examination of the bone marrow in long bones may show regeneration of the haemopoietic (red) marrow, which replaces the inactive fatty marrow (**14.2**).

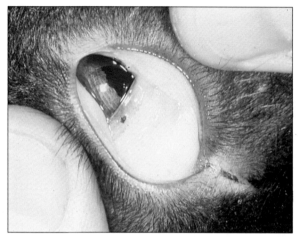

14.1 Anaemia. Acute haemonchosis (packed cell volume 0.8 litres per litre.).

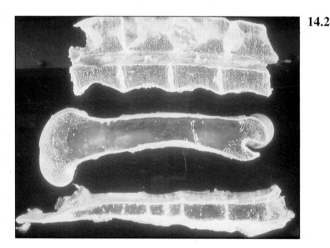

14.2 Regeneration of the bone marrow in chronic anaemia.

14.3

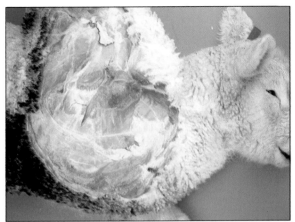

14.3 Congenital icterus in a stillborn lamb.

14.4

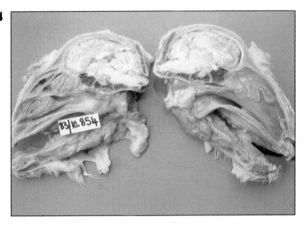

14.4 Congenital icterus. Bilirubin pigmentation of the cranial bones and non-nervous soft tissues.

14.5

14.5 Icterus. A sheep with the haemolytic crisis of copper poisoning.

Haemolytic anaemia involves destruction of circulating red cells. Common causes are infections with blood parasites (see below and Chapter 16); infectious agents, such as *Leptospira pomona* (see Chapter 11), *Clostridium perfringens* type A and *C. novyi* type D (*C. haemolyticum*); chemical poisons, such as copper and nitrates; and plant poisons, such as brassicas and bracken (see below and Chapter 17). This type of anaemia is often associated with haemoglobinuria (redwater).

Icterus

Icterus is a key indicator of haemolytic anaemia. The yellow colour in icterus is due to an increase in the bilirubin content of the blood. In haemolytic or pre-hepatic disease, unconjugated bilirubin concentrations rise due to lysis of red cells and degradation of haemoglobin, from which bilirubin is derived. The concentration of unconjugated bilirubin also rises in severe liver dysfunction; for example, in hepatogenous photosensitisation (see Chapter 10) and in some other hepatotoxicities (see Chapter 17). Icterus can also result from obstruction of the bile ducts; for example, by neoplasia. In this form, known as post-hepatic icterus, both conjugated and unconjugated bilirubin can be detected in the blood by clinical biochemistry.

A congenital form of icterus has been recognised in lambs in Australia (**14.3, 14.4**), although the cause has not been established.

In clinical icterus, visible mucous membranes are characteristically bright yellow or orange (**14.5**). At necropsy in haemolytic disease the whole carcase is yellow, the liver may be orange, while the kidneys are often dark brown or black due to the presence of haemoglobin in the tubules.

Anaplasmosis

This disease is caused by a rickettsial organism, *Anaplasma ovis*, which commonly infects both sheep and goats. These ruminants may also acquire latent infections with *A. marginale,* which normally infects cattle. The parasite is transmitted by ticks and biting flies.

Young animals acquire infections which remain latent; thus, a carrier state ensues which acts as a source of infection for newly introduced susceptible animals. After an incubation period of one-three months, such animals become depressed and develop marked pallor and icterus, without haemoglobinuria, which distinguishes the disease from babesiosis.

Incoordination and a fast heart rate may be seen. The organisms can be detected in the red cells early in the clinical disease (**14.6**).

Necropsy findings include pallor and icterus; the spleen may be enlarged and the gall bladder distended. Although in the Mediterranean region, where the disease is common, most animals become progressively anaemic and emaciated, mortality is low.

Babesiosis

Infection with the tick-borne protozoan parasites *Babesia ovis* and *B. motasi* can cause quite severe outbreaks of progressive anaemia, icterus and haemoglobinuria in sheep through destruction of erythrocytes, although many infections are inapparent. Goats are also susceptible to both species, but less commonly show icterus or haemoglobinuria. Acute deaths also occur. The parasite can be observed in erythrocytes in acute infections (**14.7**). Necropsy findings in sheep include generalised icterus, dark kidneys and splenic and hepatic enlargement.

Bracken poisoning

Bracken fern (*Pteridium aquilinum*) is distributed worldwide, and contains several toxic principles that are injurious to animals. Some are oncogenic (see Chapter 18); others cause retinal atrophy (see Chapter 12). At least one, ptaquiloside, causes severe aplasia of the bone marrow, with profound leukopenia and thrombocytopenia.

Sheep and goats generally find bracken unpalatable, and eat it only if other food is unavailable. However, in cold late spring weather, significant amounts may be ingested by sheep. The initial clinical sign is the passage of blood in the faeces and urine. An affected sheep may be followed by others, perhaps attracted by the smell of blood. High fever, melaena, and rapid respirations and heart beats may be noted.

At necropsy, there is marked pallor of the carcase, watery blood, mucosal and serosal haemorrhages (**14.8**) and septic infarcts in liver, kidneys, lungs or the rumen wall.

Goats fed experimental diets containing bracken for one–two months lost weight and developed profound leukopenia, anaemia and thrombocytopenia, although they generally recovered on resumption of a normal diet. There is some circumstantial evidence that range goats can succumb to bracken poisoning, with similar clinical indices.

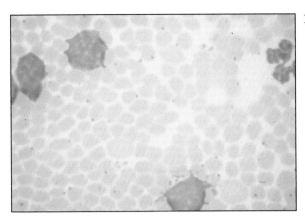

14.6

14.6 Anaplasmosis. *Anaplasma marginale* in sheep red cells. Giemsa stain.

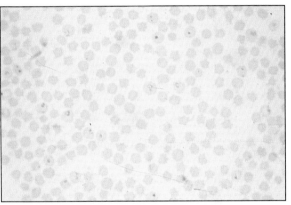

14.7

14.7 Babesiosis. *Babesia motasi* in sheep red cells. Giemsa stain.

14.8

14.8 Acute bracken poisoning. Thrombocytopenia with multiple haemorrhages.

Cow colostrum anaemia

The practice of feeding bovine colostrum to young orphan lambs and kids is sometimes followed by the development of severe anaemia within one-two weeks. This anaemia is immunologically mediated, some cows evidently secreting antibodies capable of lysing sheep or goat erythrocytes. A Coombs test will detect bovine IgG attached to the red cells of most affected young lambs.

Clinically affected lambs are weak and have marked pallor of the conjunctivae (**14.9**). They may have a packed cell volume of less than 0.10 litres per litre. At necropsy, carcases are very pale, with watery blood and flabby heart muscle. A notable finding in long bones is a cream-coloured bone marrow, consistent with haemopoietic exhaustion (**14.10**).

14.9

14.9 Cow colostrum anaemia. Compare the sclera of the anaemic lamb (left) with the normal lamb.

14.10

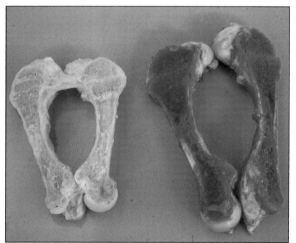

14.10 Cow colostrum anaemia. Haemopoietic exhaustion in the bone marrow (left).

Eperythrozoonosis

The rickettsial organism *Eperythrozoon ovis* can cause anaemia and mild icterus in sheep. Clinical signs are reported only rarely in goats. A seasonal incidence of infection indicates that the disease is spread by ticks and biting flies. Ring-shaped bacillary or coccobacillary forms of *Eperythrozoon* can be seen in stained blood films, attached to the surface of the red cells (**14.11**), and some may be found in the plasma, a point of differentiation from *Anaplasma* spp.

After an incubation period of about a week, the parasites multiply rapidly to outnumber the available red cells. Numbers then decline as anaemia develops. Infections are often inapparent, but infected lambs sometimes die. The anaemia and debility probably render some sheep more susceptible to other tick-borne infections, such as tick-borne fever (see below). Sick animals are anaemic and icteric, but haemoglobinuria seldom occurs. Necropsy findings include serous effusion in body cavities, pallor, moderate icterus and enlargement of the spleen. The kidneys may be abnormally pigmented due to haemosiderin deposition in the tubules (**14.12**).

14.

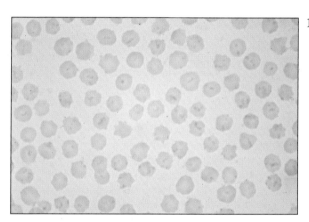

14.11 Eperythrozoonosis. *Eperythrozoon ovis* forms on sheep red cells. Giemsa stain.

14.

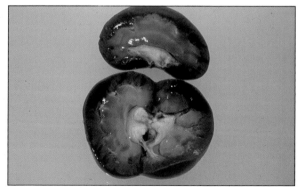

14.12 Eperythrozoonosis. Pink appearance of renal medulla due to haemosiderosis.

Rape/kale anaemia

Certain brassicas, including popular fodder crops such as rape and kale, contain alcoholic disulphides which can cause haemolytic anaemia. *S*-methyl-cysteine sulphoxide (SMCO) is the precursor of the haemolytic factor, and may be fermented in the rumen to produce dimethyl disulphide, which causes haemolysis. The severity of the anaemia is directly related to the amount of SMCO ingested.

Severe anaemia with icterus are the main clinical signs, often associated with haemoglobinuria. Examination of blood films may confirm the presence of Heinz-Ehrlich bodies in the red cells. Subclinically affected animals are unthrifty and languid. Necropsy findings include pallor and icterus, and the spleen, liver and kidneys may be dark brown due to the deposition of haemosiderin (**14.13, 14.14**).

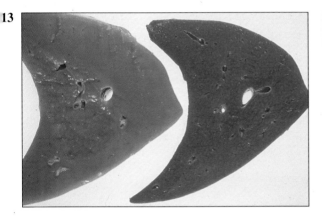

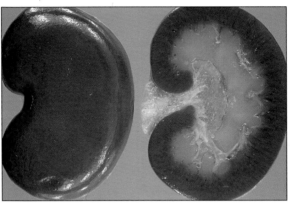

14.13 Rape/kale anaemia. Haemosiderosis of a sheep liver (right) demonstrated by the Prussian blue reaction.

14.14 Rape/kale anaemia. Haemosiderosis of a sheep kidney.

Stachybotryotoxicosis

Contamination of cereal grains or hay by the fungus *Stachybotrys alternans* causes sporadic lip fissuring, buccal and oral oedema, and progressive thrombocytopenia and leukopenia in sheep. The disease is particularly common in eastern Europe and in the Commonwealth of Independent States.

The main lesions are focal haemorrhage and necrosis in the lips, tonsillar fauces and alimentary tract mucosa (**14.15**), especially the large intestine. Surviving sheep often lose their fleece.

14.15 Stachybotryotoxicosis. Multiple haemorrhages in the intestines.

14.16

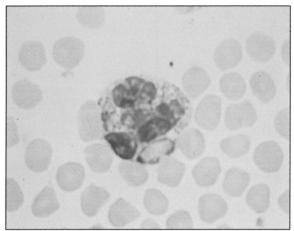

14.16 Tick-borne fever. *Cytoecetes phagocytophila* in eosinophil leukocyte. Leishman stain.

14.17

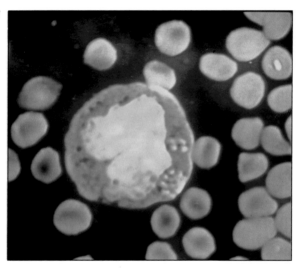

14.17 Tick-borne fever. *Cytoecetes phagocytophila* in leukocyte demonstrated by fluorescence after acridine orange staining.

Tick-borne fever

This disease of sheep and goats is caused by the rickett-sial agent *Cytoecetes phagocytophila* in Europe and *C. ovis* in India. As its name suggests, it is transmitted by ticks: *Ixodes ricinus* in Europe and *Rhipicephalus haemaphysaloides* in Asia. The vector in South Africa is unknown.

Clinical signs are vague, but initially there is a high fever which may last for several weeks. A rapid pulse, shallow respirations and depression of appetite may be noted. Animals may be unthrifty for some time, and rams may have impaired spermatogenesis for several months. Pregnant animals may abort.

The haematological changes are neutrophilia initial-ly, followed abruptly by a severe leukopenia and thrombocytopenia. Parasitaemia is recognisable during the febrile stage, the organism being easily demon-strated in the cytoplasm of granulocytes in stained blood films (**14.16, 14.17**). Necropsy findings include splenic enlargement, serosal petechiation and hydropericardium (**14.18**).

14

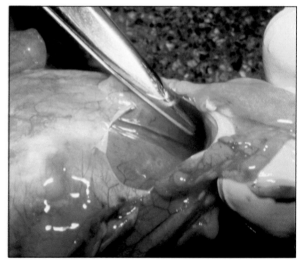

14.18 Tick-borne fever. Pericardial effusion in a young sheep.

Diseases of the lymphatic system

Caseous lymphadenitis

This important disease of sheep and goats is caused by *Corynebacterium pseudotuberculosis (C. ovis)*, and results in considerable economic loss due to condemnation of carcasses at the abattoir. The disease is acquired by infection of wounds, either by soil contaminated by the organism or mechanically (by shearing, for example).

In the superficial form of the disease, there is marked enlargement of the superficial lymph nodes **(14.19–14.21)** due to the progressive development of abscesses, which can develop at any site. The precrural and prescapular nodes are most commonly affected in sheep, and the parotid and submandibular nodes in goats. Large abscesses are the result of coalescence of microabscesses, with accumulation of leukocytes, caseation and encapsulation followed by further sequences of these changes peripherally so that eventually the abscesses have a lamellated 'onion-ring' appearance in section **(14.22)**. Abscesses often rupture, leaving fistulae discharging through the skin and allowing the organism to contaminate the environment. Introduction of an infected animal into a susceptible group can thus cause rapid spread of the disease, although the developing abscesses are not apparent for 2–6 months.

The visceral form of the disease is insidious, with gradual loss of condition leading to emaciation. Lesions can be found in a variety of sites, including lungs, kidneys and even the spine (see Chapters 5, 7 and 11). The disease has zoonotic potential in that some human cases have been reported.

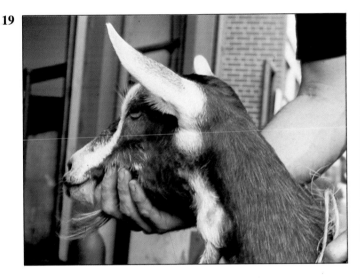

14.19 Caseous lymphadenitis. Abscesses in a parotid lymph node and another node near the base of a wattle.

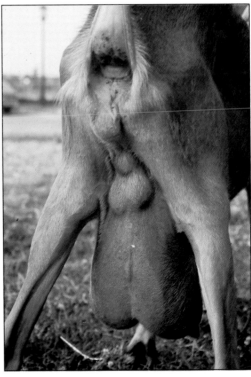

14.20 Caseous lymphadenitis. Supramammary lymph nodes and other lymph nodes in the perineum are infected.

14.21

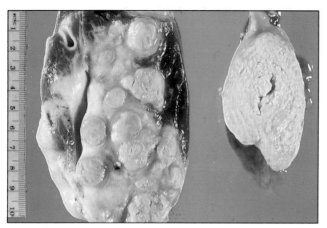

14.21 Caseous lymphadenitis. Lesions in sheep liver and lymph node.

14

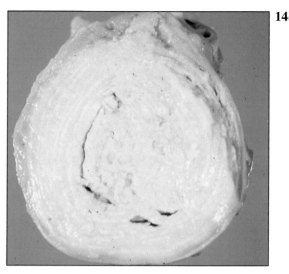

14.22 Caseous lymphadenitis. Lamellated lesion in a goat lymph node.

Coccidiosis of lymph nodes

The mesenteric lymph nodes are commonly infected with endogenous stages of various *Eimeria* species, mainly meronts or microgametocytes (**14.23**). These stages eventually collapse and become microgranulomas. They have no clinical significance.

14.23

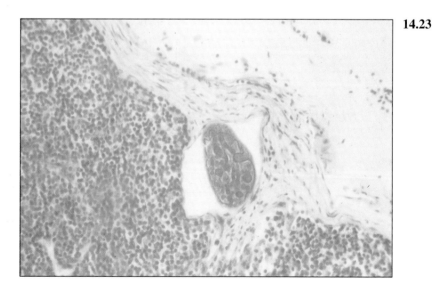

14.23 Coccidial stage in a sheep mesenteric lymph node. Haematoxylin and eosin.

Lymphosarcoma and thymic lymphosarcoma

Lymphosarcoma is probably the most common neoplasm that affects young sheep; it can also occur in older sheep and goats. Affected animals are unthrifty, and some have symmetrically enlarged superficial lymph nodes (**14.24**, **14.25**). By the time this enlargement is obvious, the spleen and other internal lymphatic sites are also involved, and the neoplastic process may have spread to non-lymphatic organs such as liver, kidneys, heart, bone marrow and alimentary tract (**14.26–14.29**) and lungs (see **7.58**). A localised form causing nodular skin tumours (**14.30**) can occur. Some goats have extensive involvement of internal organs without enlargement of peripheral lymph nodes.

In flocks where multiple cases of lymphosarcoma

14.24 Lymphosarcoma. A goat with generalised disease.

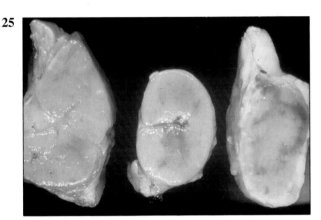

14.25 Lymphosarcoma. Neoplastic enlargement of prescapular nodes compared with a normal node (centre).

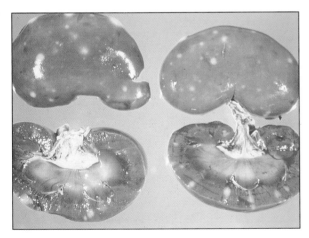

14.27 Lymphosarcoma. Kidney secondaries in a goat.

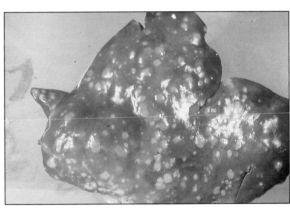

14.26 Lymphosarcoma. Liver lesions in a sheep.

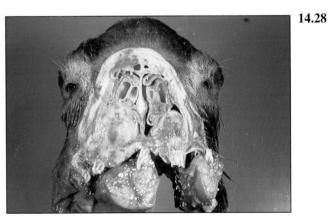

14.28 Lymphosarcoma. Neoplasia has spread to the jaws in this goat.

14.29

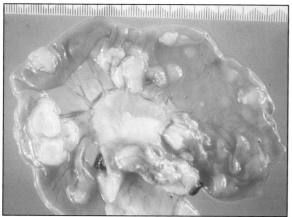

14.29 Lymphosarcoma. Lesions in the intestine of a sheep.

14.30 Lymphosarcoma. Sheep with skin nodules of lymphosarcoma.

14.31

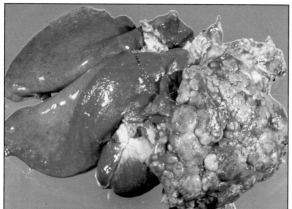

14.31 Thymic lymphosarcoma in a sheep.

occur, many animals may have a marked lymphocytosis, and there is growing evidence that a viral agent might be responsible.

Thymic lymphoma or lymphosarcoma is a sporadic disease characterised by massive enlargement of the thymus (**14.31, 14.32**) and progressive emaciation. Care must be taken not to interpret the physiological prominence of the thymus in young goats (**14.33**) as thymic lymphoma.

14.32

14.32 Thymic lymphosarcoma. Massive tumour in an old goat.

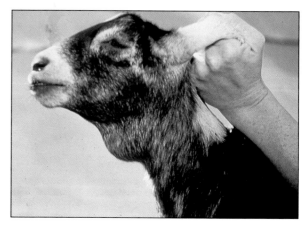

14.33 Normal thymus development in a young Nubian goat.